AF131136

Synthesis Lectures on Engineering, Science, and Technology

The focus of this series is general topics, and applications about, and for, engineers and scientists on a wide array of applications, methods and advances. Most titles cover subjects such as professional development, education, and study skills, as well as basic introductory undergraduate material and other topics appropriate for a broader and less technical audience.

Felix Staudigl • Rainer Leupers

Towards Trustworthy Neuromorphic Computing

An Analysis of Hardware Security and Reliability Risks

Felix Staudigl (iD)
RWTH Aachen University
Aachen, Germany

Rainer Leupers
RWTH Aachen University
Aachen, Germany

ISSN 2690-0300   ISSN 2690-0327   (electronic)
Synthesis Lectures on Engineering, Science, and Technology
ISBN 978-3-032-09585-5   ISBN 978-3-032-09586-2   (eBook)
https://doi.org/10.1007/978-3-032-09586-2

© The Editor(s) (if applicable) and The Author(s), under exclusive license to Springer Nature Switzerland AG
2026

This work is subject to copyright. All rights are solely and exclusively licensed by the Publisher, whether the
whole or part of the material is concerned, specifically the rights of translation, reprinting, reuse of illustrations,
recitation, broadcasting, reproduction on microfilms or in any other physical way, and transmission or information
storage and retrieval, electronic adaptation, computer software, or by similar or dissimilar methodology now
known or hereafter developed.
The use of general descriptive names, registered names, trademarks, service marks, etc. in this publication does
not imply, even in the absence of a specific statement, that such names are exempt from the relevant protective
laws and regulations and therefore free for general use.
The publisher, the authors and the editors are safe to assume that the advice and information in this book are
believed to be true and accurate at the date of publication. Neither the publisher nor the authors or the editors
give a warranty, expressed or implied, with respect to the material contained herein or for any errors or omissions
that may have been made. The publisher remains neutral with regard to jurisdictional claims in published maps
and institutional affiliations.

This Springer imprint is published by the registered company Springer Nature Switzerland AG
The registered company address is: Gewerbestrasse 11, 6330 Cham, Switzerland

If disposing of this product, please recycle the paper.

Declarations

Competing Interests The studies summarized in this book were funded by Federal Ministry of Education and Research (BMBF, Germany) in the project NEUROTEC I/II (16ES1134,16ME0399). Furthermore, the authors have no conflicts of interest to declare that are relevant to the content of this chapter.

Acknowledgments

We thank everyone who contributed their time, effort, and insight to this book. This work would not exist without the support, feedback, and patience of all who helped along the way.

Contents

Acronyms

1T1R	1-Transistor 1-Resistor
ADC	Analog-to-Digital Converter
ASIC	Application-Specific Integrated Circuit
AES	Advanced Encryption Standard
API	Application Programming Interface
BNN	Binary Neural Network
BL	Bit Line
BEOL	Back End Of Line
BE	Bottom Electrode
BLC	Bit Line Controller
CIM	Computing-in-Memory
CMOS	Complementary Metal-Oxide Semiconductor
C2C	Cycle-to-Cycle
CRT	Chinese Remainder Theorem
CDF	Cumulative Distribution Function
DAC	Digital-to-Analog Converter
DRAM	Dynamic Random-Access Memory
D2D	Device-to-Device
DFS	Design For Security
DNN	Deep Neural Network
DMA	Direct Memory Access
DoS	Denial-of-Service
eNVM	Emerging Non-Volatile Memory
ECC	Error Correcting Code
ENTT	Emerging NVM-based Trojan Trigger
EFI	Enhanced Fault Injection Attack
FLIM	Faulty Logic-in-Memory
FEOL	Front End Of Line
FPGA	Field-Programmable Gate Array
FCM	Fast Crossbar Model
FMC	FPGA Mezzanine Card

FTJ	Ferroelectric Tunnel Junction
GND	Ground
GLC	Gate Line Controller
GPDK	Generic Process Design Kit
GPIO	General-Purpose Input/Output
GUI	Graphical User Interface
HRS	High Resistive State
HAL	Hardware Abstraction Layer
IC	Integrated Circuit
IMPLY	Memristor-Based Material Implication
IP	Intellectual Property
IMP	Material Implication
IFV	Incremental Form and Verify
ISPVA	Incremental Step Pulse with Verify Algorithm
JTAG	Joint Test Action Group
LIM	Logic-in-Memory
LRS	Low Resistive State
LLC	Last Level Cache
LSB	Least Significant Bit
MAC	Multiply–Accumulate
MAGIC	Memristor-Aided Logic
MRL	Memristor Ratioed Logic
mCAT	Memristor Characterization And Testing
MIM	Metal-Insulator-Metal
NBB	NeuroBreakoutBoard
NMOS	N-Type Metal–Oxide Semiconductor
NN	Neural Network
PUF	Physical Unclonable Function
PCB	Printed Circuit Board
PC	Popcount
PCM	Phase Change Memory
QFP	Quad Flat Package
ReRAM	Resistive Random-Access Memory
RSA	Rivest–Shamir–Adleman
RNG	Random Number Generation
SRAM	Static Random-Access Memory
STT-RAM	Spin-Transfer Torque Random-Access Memory
SMU	Source-Meter Unit
SPI	Serial Peripheral Interface
SWD	Serial Wire Debug
TIA	Transimpedance Amplifier
TE	Top Electrode

UART	Universal Asynchronous Receiver/Transmitter
VCM	Valence Change Material
VADER	Variation-oriented Adversarial Attack
WL	Word Line
WLC	World Line Controller

Introduction

1

The von Neumann bottleneck has long been recognized as a significant impediment to the performance of conventional computing systems. This bottleneck arises from the strict separation between memory and processing units, necessitating frequent data transfers and resulting in suboptimal system performance and energy efficiency. To overcome this challenge, emerging computing paradigms, such as Computing-in-Memory (CIM), have garnered considerable attention. By shifting computational operations inside the memory, CIM endeavors to mitigate the limitations imposed by the von Neumann bottleneck. The concept of relocating computational operations to memory is derived from the mammalian brain, giving rise to the term *"neuromorphic computing."*

Memristors, initially postulated as the fourth fundamental circuit element by Leon Chua [1] in 1971, serve as the foundational building blocks for neuromorphic computing systems. Leveraging the resistive switching characteristics of memristors, these devices offer promising advantages, including high density, non-volatility, and low static power consumption. Based on these unique characteristics, memristors allow the implementation of two flavors of CIM: analog CIM and Logic-in-Memory (LIM).

Analog CIM leverages the parallelism inherent in memristive devices to perform computational operations directly within the memory arrays. This approach offers the potential for higher throughput and overall computing performance compared to conventional system architectures but requires the use of Analog-to-Digital Converters (ADCs) and Digital-to-Analog Converters (DACs). On the other hand, LIM takes a different approach by implementing binary logic gates directly within the memristive crossbar arrays. This eliminates the need for ADCs and DACs, simplifying the system architecture but offers lower computing performance.

However, the reliability of memristors remains a critical concern, exerting a substantial influence on the overall reliability of CIM architectures. For analog CIM, the variability of

© The Author(s), under exclusive license to Springer Nature Switzerland AG 2026

F. Staudigl, R. Leupers, *Towards Trustworthy Neuromorphic Computing*,
Synthesis Lectures on Engineering, Science, and Technology,
https://doi.org/10.1007/978-3-032-09586-2_1

memristive devices, caused by fabrication process variations and aging effects, introduces uncertainties and affects accuracy. LIM may offer a more robust solution compared to analog CIM, as it operates in the digital domain, which is less affected by the inherent variability of memristive devices.

Alongside performance limitations, conventional computing systems also suffer from inherent hardware security vulnerabilities. Hardware security focuses on identifying specific design characteristics that may be exploited to gain unauthorized control over the entire system [6]. Due to the rigid structure of Integrated Circuits (ICs), these vulnerabilities pose considerable challenges in terms of remediation, persisting throughout the entire operational lifespan of the IC. The emergence of hardware-enabled attacks, including notable examples such as Rowhammer [2], Spectre [3], and Meltdown [4], has underscored the potential for significant disruptions to critical computing infrastructure. Consequently, the investigation of hardware security vulnerabilities and the development of corresponding countermeasures have become a vital area of research [5]. These vulnerabilities pose a grave risk to the confidentiality and integrity of sensitive data, posing an interesting question: *Is neuromorphic computing susceptible to hardware security attacks?*

Reliability issues and hardware security risks are critical barriers in the development of neuromorphic systems that use memristive devices. This book examines these two concerns, with a focus on computing-in-memory (CIM) architectures built on memristors. The aim is to understand the underlying causes of reliability and security problems in these systems to propose ways to reduce their impact. The results contribute to building more secure and dependable computing systems for future use. The main contributions are as follows:

Fault Injection Platform The implementation of the first fault injection framework for LIM operations, capable of investigating reliability from the memristor level to actual full-fledged workloads. The framework includes a memristor-level fault injection simulator named X-Fault and an operational-level simulator called Faulty Logic-in-Memory (FLIM) . The former excels at emulating the impact of faults on individual logic gates with high precision, while the latter offers exceptional simulation speed for executing realistic workloads using abstracted fault models. Together, these simulators allow for a comparison of logic families in terms of fault resilience and enable an investigation at the application level to understand which parameters most significantly influence potential workloads. While both simulators use fault models from the literature, verifying our simulation results with actual hardware is beyond the scope of this work.

NeuroHammer A novel hardware security attack termed NeuroHammer, which threatens the integrity of the entire neuromorphic system. This attack specifically targets the memristive crossbar arrays utilized in neuromorphic computing systems, intentionally inducing bit-flip faults to undermine the foundational principle of modern computing systems–memory separation. By exploiting the distinct properties of Resistive Random-

Access Memory (ReRAM) to alter the switching kinetics through thermal crosstalk, NeuroHammer unveils an attack surface, similar to the Rowhammer attack in Dynamic Random-Access Memory (DRAM). We provide a comprehensive case study showcasing the profound impact of NeuroHammer by leaking an Rivest–Shamir–Adleman (RSA) key from a computing system using memristive memory.

NeuroBreakoutBoard (NBB) The design and implementation of the NBB , a versatile and adaptable instrumentation platform designed to investigate the characteristics of memristive devices at the device, crossbar, and operational levels. Equipped with custom-designed signal generation and sensing circuitry, the NBB enables precise control over memristive cell programming and supports the execution of both analog CIM and LIM operations.

The remainder of this book is organized as follows. Chapter 2 provides an overview of memristive memory technology, the CIM paradigm, and reliability concerns associated with memristive devices. The related literature is discussed in Chap. 3. Chapter 4 introduces the fault injection framework, along with two comprehensive case studies investigating the fault resilience of logic families and Binary Neural Networks (BNNs). Following this, Chap. 5 presents NeuroHammer, accompanied by a case study illustrating its potential risks through the leakage of an RSA key. Chapter 6 explores the design and implementation of the NBB , followed by a characterization of a commercially available ReRAM technology node. Lastly, Chap. 7 concludes the book and outlines future research directions.

References

1. Chua, L.: Memristor-the missing circuit element. IEEE Trans. Circuit Theory **18**(5), 507–519 (1971). https://doi.org/10.1109/tct.1971.1083337
2. Kim, Y., Daly, R., Kim, J., Fallin, C., Lee, J.H., Lee, D., Wilkerson, C., Lai, K., Mutlu, O.: Flipping bits in memory without accessing them. ACM SIGARCH Comput. Archit. News **42**(3), 361–372 (2014). https://doi.org/10.1145/2678373.2665726
3. Kocher, P., Horn, J., Fogh, A., Genkin, D., Gruss, D., Haas, W., Hamburg, M., Lipp, M., Mangard, S., Prescher, T., Schwarz, M., Yarom, Y.: Spectre attacks: exploiting speculative execution. In: IEEE Symposium on Security AND Privacy (S&P). IEEE (2019). https://doi.org/10.1109/sp.2019.00002
4. Lipp, M., Schwarz, M., Gruss, D., Prescher, T., Haas, W., Fogh, A., Horn, J., Mangard, S., Kocher, P., Genkin, D., Yarom, Y., Hamburg, M.: Meltdown: reading kernel memory from user space. In: USENIX Security Symposium (USENIX Security 18), pp. 973–990. USENIX Association, Baltimore (2018). https://www.usenix.org/conference/usenixsecurity18/presentation/lipp
5. Rai, S., Garg, S., Pilato, C., Herdt, V., Moussavi, E., Sisejkovic, D., Karri, R., Drechsler, R., Merchant, F., Kumar, A.: Vertical IP protection of the next-generation devices: quo vadis? In: Design, Automation & Test in Europe Conference & Exhibition (DATE). IEEE (2021). https://doi.org/10.23919/date51398.2021.9474132
6. Seaborn, M., Dullien, T.: Exploiting the DRAM rowhammer bug to gain kernel privileges (2015). http://googleprojectzero.blogspot.com.tr/2015/03/exploiting-dram-rowhammer-bug-to-gain.html

Background

2

The objective of this chapter is to provide the essential background information that will facilitate a comprehensive understanding of the contributions made in this book. Section 2.1 explores the working principles of Emerging Non-Volatiles Memories (eNVMs). The CIM paradigm is explained in Sect. 2.2. Moreover, Sect. 2.3 discusses the reliability aspects associated with memristor-based memories. The hardware security of both conventional and emerging computing systems is detailed in Sect. 2.4. Finally, Sect. 2.5 marks the conclusion of this chapter.

2.1   Emerging Non-Volatile Memories (eNVMs)

Memristive devices serve as fundamental building block of eNVMs. The following section introduces the working principles of these devices and discusses typical memory structures.

2.1.1   Memristive Devices

The data storage in memristive devices hinges upon altering resistance by applying specific voltage pulses across the device terminals. Varied resistive states are achievable based on the polarity, amplitude, and pulse duration, increasing memory density and positioning this technology at the forefront of memory applications and in-memory computing paradigms [85]. In the case of binary switching devices, resistive states are identified as the High Resistive State (HRS) and Low Resistive State (LRS). Memristive devices are categorized according to the underlying resistance switching mechanism.

© The Author(s), under exclusive license to Springer Nature Switzerland AG 2026   5
F. Staudigl, R. Leupers, *Towards Trustworthy Neuromorphic Computing*,
Synthesis Lectures on Engineering, Science, and Technology,
https://doi.org/10.1007/978-3-032-09586-2_2

Diverse mechanisms are employed by various types of memristive devices. Phase Change Memories (PCMs) transition from a non-conducting phase (amorphous) to a conducting phase (crystalline) via Joule heating processes [30]. Spin-Transfer Torque Random-Access Memories (STT-RAMs) utilize spin-polarized currents to switch the magnetization of the free layer in magnetic tunnel junctions, limiting tunneling current [77]. In Ferroelectric Tunnel Junctions (FTJs), the transport of electrons is modulated by the field-induced polarization switching of a ferroelectric layer [18].

However, this book centers on Resistive Random-Access Memories (ReRAMs), which modulate resistance through the redistribution of ionic defects [94]. Specifically, the Valence Change Material (VCM) is employed, typically utilizing oxygen vacancies as mobile defects to modify local conductivity. This alteration modifies the electrostatic barriers at the metal/oxide interfaces by changing the width of the depletion layer [29, 95, 102]. This redistribution of ionic defects within a specific filamentary region allows for distinct resistive states. For instance, a high concentration of oxygen vacancies at the metal interface's disc region, known as the active electrode interface, places the device in the LRS [70]. Conversely, a low concentration of oxygen vacancies at this interface results in the device being in the HRS. The transition between these states, known as the SET transition (HRS to LRS) or RESET transition (LRS to HRS), is achieved by applying specific voltage polarities to the active electrode. The movement of oxygen vacancies, driven by their positive charge, governs these transitions. Joule heating can expedite ion migration, facilitating highly nonlinear switching dynamics, stable read operations at low voltages, and faster switching at slightly higher voltages [71, 97]. Theoretical models by Menzel et al. [71, 72] offer insights into relevant model parameters associated with these dynamics. Filamentary VCM cells are preferred for emerging computing paradigms due to their compatibility with fabrication processes and Complementary Metal-Oxide Semiconductor (CMOS) technology. Table 2.1 provides an overview of commercial and academic prototypes utilizing ReRAMs. Promising oxide materials like HfO_2 and Ta_2O_5, used by companies such as Panasonic and TSMC for ReRAM macros, highlight their in-memory computing capabilities [8, 14, 40, 54, 60, 92, 98, 105].

2.1.2 Memristive Crossbar Structures

Crossbar structures stand out as the primary choice for achieving high-density memristive memories. At their core, these structures interconnect memristive cells through vertical Bit Line (BL) and horizontal Word Line (WL), as depicted in Fig. 2.1a. To program a single memristive cell, a high voltage V_{write} is applied to the corresponding WL, while the corresponding BL is set to Ground (GND). Ensuring the non-selected cells remain protected against unintended programming is essential. All other lines are set to $V_{write}/2$ to adhere to the "$V/2$ scheme" [61]. Ideally, a voltage drop of V_{write} occurs across the selected device, while the non-selected devices along the WL and BL experience an absolute value of $|V_{write}/2|$. However, the ideal scenario, disregarding line resistances and

Table 2.1 Summary of commercial and academic prototypes using ReRAMs for memory and computing applications [26]

Institution	Node	Stack	Capacity	Cell	Endurance	Ref.
Crossbar Inc.	180 nm			1R		[44]
	45 nm			1T1R		[20]
Sony	180 nm	Cu CBRAM	4 MB	2T 1S1R	10^7	[76]
		CuTe CBRAM		1T1R		[104]
TSMC	40 nm	HfOx	11 MB	1T1R	10^4	[14]
	22 nm		13.5 MB			[15]
Weebit Nano	28 nm	HfO2	16 kB	1T1R	10^5	[32]
Tsinghua Uni.	130 nm	HfOx	16 MB	1T1R	10^6	[13]
Intel	22 nm	Cu CBRAM	4 MB	1T1R		[42]
Panasonic	40 nm	TaOx	1 Mbit		10^5	[105]
HP Labs		HfO2	16 kB	1T1R		[59]
IHP	130 nm	HfO2	4 kbit	1T1R		[96]

capacitances, introduces parasitic sneak path currents. These currents not only impact the read operation's accuracy but also limit array size, induce undesired switching events, and increase power dissipation [62].

Hence, more sophisticated crossbar structures emerged designed to mitigate sneak path currents in passive crossbar arrays. Intel's 3D Xpoint memory, for instance, integrates threshold switches in series with the memristive devices, effectively reducing sneak path currents. Although this design offers high memory density, it necessitates increased switching voltages [41]. Another approach employs active crossbar arrays incorporating transistors as selectors. However, as transistors are three-terminal devices, introducing an additional line connecting the transistor gates reduces memory density. Figure 2.1b–d illustrates three distinct 1-Transistor 1-Resistor (1T1R) topologies–typical 1T1R array, vertical 1T1R array, and pseudo-crossbar array–each characterized by different arrangements involving exclusively horizontal and vertical lines [83]. In the following, the writing schemes for the active crossbar structures are detailed corresponding to Fig. 2.1:

Typical 1T1R array: To write a cell (n,m), all transistors of the n-th line are selected by applying the respective gate voltage V_{Gate}, while all other select lines are set to GND. Naturally, only the devices along the selected row can be programmed. Due to the WLs and BLs are arranged in parallel, the device (n,m) can be programmed by applying voltages to the m-th WL and the m-th BL. All other WLs and BLs are set to 0 V, resulting in a voltage drop solely across the desired device (n,m) (see Fig. 2.1b).

Vertical 1T1R array: In the vertical 1T1R configuration, the transistor gates within a column share the same select line. Consequently, to program a cell (n,m), the m-th select line is set to V_{Gate}, the n-th WL to V_{Write}, the m-th BL is set to GND, and the other WLs are set to same potential as the m-th BL.

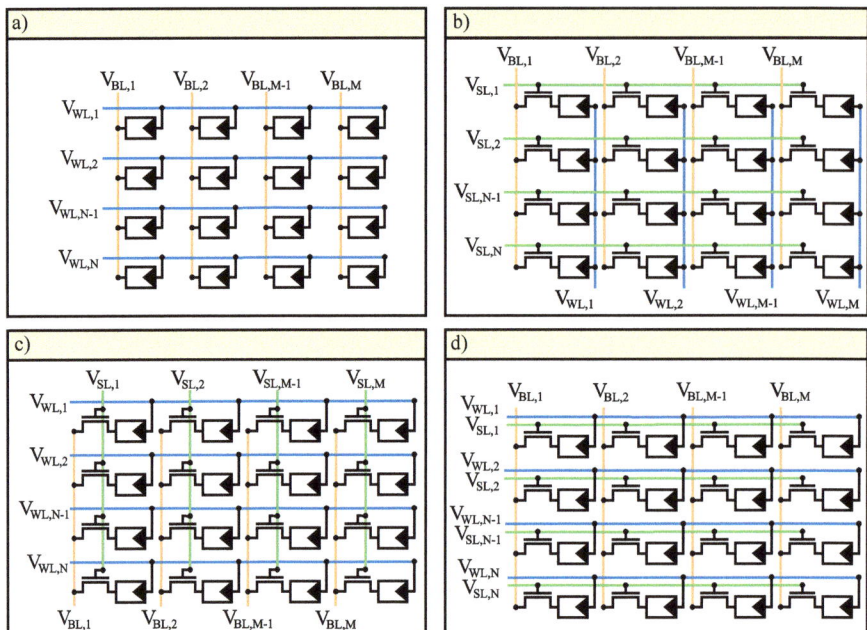

Fig. 2.1 Overview of memristive crossbar structures: (**a**) passive crossbar, (**b**) typical crossbar, (**c**) vertical crossbar, and (**d**) pseudocrossbar

Pseudo 1T1R array: In the pseudo-crossbar array, the select line and the word line are parallel to each other. To program cell (n,m), the n-th select line is set to a high potential, the n-th WL is set to V_{Write}, the m-th BL is set to GND, and the remaining BLs need to be set to V_{Write} to prevent switching in other cells of the same row.

While the described writing schemes summarized in Table 2.2 effectively mitigate sneak path currents, active crossbar structures also suffer from shortcomings. For instance, the utilization of both negative and positive write voltages could cause leakage currents through the parasitic diodes to the transistor bulk.

Moreover, the voltage drop over WLs and BLs (IR drop) has the potential to impact the read and program pulses significantly. Irrespective of the 1T1R configurations, one terminal of each memristive cell is linked through a shared line with other devices in a row or column. Consequently, when a selected transistor permits the flow of current, it induces a potential shift at the shared line due to the IR drop. The extent of the IR drop is influenced by various factors, such as the length of the current path between the WL drivers and the BL drivers. This length, often termed the critical length, fluctuates depending on the position of the selected cell within the pseudo and the vertical 1T1R array. However, it remains constant regardless of the cell's placement in the typical 1T1R array, as shown in Fig. 2.1b–d.

Table 2.2 Writing schemes for the SET/RESET operation on a typical, vertical, pseudo, and passive crossbar array

Type	SET		RESET	
Typical	$V_{\text{WL},m} = V_{\text{SET}}$ $V_{\text{SL},n} = V_{\text{GATE}}$ $V_{\text{BL},[0,M]} = \text{GND}$	$V_{\text{WL},[0,M]\backslash m} = \text{GND}$ $V_{\text{SL},[0,N]\backslash n} = \text{GND}$	$V_{\text{WL},m} = V_{\text{RESET}}$ $V_{\text{SL},n} = V_{\text{GATE}}$ $V_{\text{BL},[0,M]} = \text{GND}$	$V_{\text{WL},[0,M]\backslash m} = \text{GND}$ $V_{\text{SL},[0,N]\backslash n} = \text{GND}$
Vertical	$V_{\text{WL},n} = V_{\text{SET}}$ $V_{\text{SL},m} = V_{\text{GATE}}$ $V_{\text{BL},[0,M]} = \text{GND}$	$V_{\text{WL},[0,N]\backslash n} = \text{GND}$ $V_{\text{SL},[0,M]\backslash m} = \text{GND}$	$V_{\text{WL},n} = V_{\text{RESET}}$ $V_{\text{SL},m} = V_{\text{GATE}}$ $V_{\text{BL},[0,M]} = \text{GND}$	$V_{\text{WL},[0,N]\backslash n} = \text{GND}$ $V_{\text{SL},[0,M]\backslash m} = \text{GND}$
Pseudo	$V_{\text{WL},n} = V_{\text{SET}}$ $V_{\text{SL},n} = V_{\text{GATE}}$ $V_{\text{BL},m} = \text{GND}$	$V_{\text{WL},[0,N]\backslash n} = \text{GND}$ $V_{\text{SL},[0,N]\backslash n} = \text{GND}$ $V_{\text{BL},[0,M]\backslash m} = V_{\text{SET}}$	$V_{\text{WL},n} = V_{\text{RESET}}$ $V_{\text{SL},n} = V_{\text{GATE}}$ $V_{\text{BL},m} = \text{GND}$	$V_{\text{WL},[0,N]\backslash n} = \text{GND}$ $V_{\text{SL},[0,N]\backslash n} = \text{GND}$ $V_{\text{BL},[0,M]\backslash m} = V_{\text{RESET}}$
Passive	$V_{\text{WL},m} = V_{\text{SET}}$ $V_{\text{BL},n} = \text{GND}$	$V_{\text{WL},[0,M]\backslash m} = \frac{V_{\text{SET}}}{2}$ $V_{\text{BL},[0,N]\backslash n} = \frac{V_{\text{SET}}}{2}$	$V_{\text{WL},m} = V_{\text{RESET}}$ $V_{\text{BL},n} = \text{GND}$	$V_{\text{WL},[0,M]\backslash m} = \frac{V_{\text{RESET}}}{2}$ $V_{\text{BL},[0,N]\backslash n} = \frac{V_{\text{RESET}}}{2}$

Similarly, the leakage currents from the transistor might result in an unintended voltage drop over an unselected cell if a potential is applied between the WL and BL. These leakage currents are anticipated to emerge in more advanced technology nodes, posing a threat to the reliability of active crossbar structures.

2.2　Computing-in-Memory (CIM)

Besides the utilization of memristive crossbar arrays as memories, these structures are capable of facilitating CIM operations. In general, CIM can be realized in two different flavors: analog CIM and Logic-in-Memory (LIM).

2.2.1　Analog Computing-in-Memory (CIM)

Analog CIM harnesses the continuous resistance values of memristors to conduct Multiply–Accumulate (MAC) operations in the analog domain. As depicted in Fig. 2.2a, the operational principle of analog CIM involves the Digital-to-Analog Converter (DAC) translating the binary input vector into respective voltages and feeding them to the crossbar's rows. According to Ohm's and Kirchhoff's law, the resulting current of column c is defined as:

$$i_{c,\text{res}} = \sum_{r=1}^{R} G_{r,c} V_r \tag{2.1}$$

Here, R signifies the number of rows, $G_{r,c}$ indicates the conductance of the memristor in row r and column c, and V_r stands for the input voltage applied to row r. Subsequently,

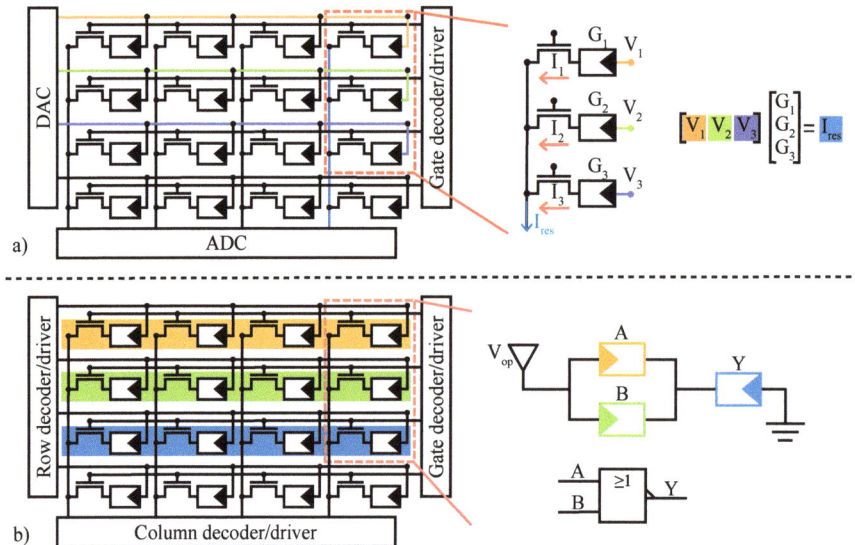

Fig. 2.2 Overview of CIM flavors: (**a**) analog CIM and (**b**) LIM

the ADC translates the output current into a digital value. Analog CIM facilitates the computation of MAC operations in a massively parallel manner, providing high precision and low latency, which is beneficial for applications such as machine learning and signal processing. However, it involves complex peripheral circuitry, notably utilizing ADCs/DACs, which are known for their considerable silicon footprint, increased energy consumption, and latency limitations [89]. Moreover, the variability of the memristors directly impacts $i_{c,res}$, leading to a reduction in the accuracy of the executed application [65].

2.2.2   Logic-in-Memory (LIM)

On the contrary, LIM leverages memristive crossbar arrays in a binary manner to execute logic operations. Numerous logic families have been proposed to integrate logic gates within memristive crossbar arrays [2, 5, 7, 34, 55, 57, 73, 99, 106]. Most of these logic families require additional CMOS circuitry, such as a CMOS inverter [52], to implement logic gates. However, stateful logic represents a group of logic families that aims to conduct logic gate operations entirely within memristive crossbar arrays, eliminating the need for complex external circuitry [37]. Stateful logic gates encode their inputs and outputs in the form of resistance. During computation, the result is directly stored into the output memory cell without data being transferred outside the memory array [79]. Subsequently, the Memristor-Based Material Implication (IMPLY) [51], Memristor-Aided

Table 2.3 Implementation details for MAGIC, IMPLY, and MRL [25, 51, 53]

	Circuit	Logic gate	Input/output encoding
MAGIC		NOR	Resistance/Resistance
IMPLY		IMPLY	Resistance/Resistance
MRL		NOR	Resistance/Voltage

Logic (MAGIC) [53], and Memristor Ratioed Logic (MRL) logic families are elaborated upon, as they represent the most notable logic families.

Memristor-Based Material Implication (IMPLY) The IMPLY logic family features a single logic gate known as the Material Implication (IMP) gate, as shown in Table 2.3 [7]. Together with the FALSE gate, which consistently yields zero, the IMPLY gate constitutes a functionally complete set. In its standard configuration, the IMPLY gate consists of two memristors, denoted as P and Q, linked to a resistor R_G with $R_{ON} < R_G < R_{OFF}$. The resulting output is written in the memristor Q based on the initial resistances p and q of the two memristors. To calculate the output of the logic gate, distinct voltages are applied to the input memristors. The voltage V_{SET} is directed to Q, while V_{COND} is connected to P, ensuring that $|V_{COND}| < |V_{SET}|$. The IMPLY logic family allows the execution of an arbitrary Boolean function using only $n + 3$ memristors, employing either FALSE operations with a single memristor or IMPLY operations with two memristors in a sequential manner [55]. Additionally, the memristive IMPLY gate can be incorporated into a memristive crossbar array where P, Q, and the necessary resistor R_G are connected via the same BL. Both voltages, V_{SET} and V_{COND}, are supplied through their respective WLs [51].

Memristor-Aided Logic (MAGIC) The IMPLY logic family's broader utility faces limitations due to the necessary external resistance R_G and the inherent deletion of input values during computation. To address these issues, the MAGIC logic family offers a functionally complete NOR gate. Table 2.3 illustrates the fundamental structure of the MAGIC NOR gate. The MAGIC NOR gate comprises two input memristors, in_1 and in_2, along with a dedicated output memristor out that retains the result value post-computation. Unlike the IMPLY gate, the MAGIC NOR gate preserves the input values. Upon initializing in_1, in_2, and out, an operational voltage V_{Op} is applied across the logic gate. As a result, the output memristor changes its internal state based on the resistances of the input memristors. The operational voltage must adhere to the constraint:

$$2V_{T,OFF} < V_O < \min\left[\frac{R_{OFF}}{2R_{ON}}V_{T,OFF}, |V_{T,ON}|\right] \quad (2.2)$$

Here, $V_{T,OFF}/V_{T,ON}$ denotes the voltage threshold of the memristor, and R_{OFF}/R_{ON} signifies the resistances for logical zero and logical one. By implementing IMPLY/MAGIC gates within a crossbar structure, these logic gates can operate in parallel by applying the operation voltage V_{Op} to the appropriate input rows. Figure 2.2b illustrates MAGIC NOR gates embedded in the crossbar structure, where each column is equipped with three memristive devices to form a NOR gate. In this example, applying V_{Op} to the first and second rows enables the execution of four parallel NOR operations.

Memristive Ratioed Logic (MRL) Given that both MAGIC and IMPLY logic families encode Boolean states in the form of resistance, they are inherently prone to state-drift and variability effects of memristive devices. MRL adopts a different methodology by encoding the output in a voltage level between V_{REF} and V_{DD}. Consequently, the result of the computation is not directly stored within the crossbar array, positioning this logic family as near-memory computing. In general, this logic family employs a NOR gate configured with a voltage divider across two memristive devices, where the input values are determined by the resistances of these devices. A logical 1 is represented by R_{HRS}, and a logical 0 is denoted by R_{LRS}, with a computing voltage of V_{REF}. For instance, when the input is {0, 0}, both memristive devices are set to LRS, leading to an output voltage of $V_{out} \approx V_{REF}$, signifying a logical 1. Conversely, for any other input combination, V_{out} is reduced to below $\frac{V_{DD}-V_{REF}}{2} + V_{REF}$, which is interpreted as a logical 0 [21, 22, 25].

2.3   Reliability Aspects of ReRAMs

Notwithstanding their potential to enable novel computing paradigms, memristors face susceptibility to faults attributed to an immature manufacturing process and limited endurance [67]. Consequently, this section serves to introduce crucial terminology, describe relevant fault models, and explore fault injection methodologies concerning ReRAMs.

2.3.1   Terminology

In general, a system may deviate from its intended operation due to various *factors of dependability*, which describe the origins and consequences of system malfunctions as follows [107]:

Definition 2.1 A **fault** is a physical defect, imperfection, or flaw within hardware or software.

Definition 2.2 An **error** represents a departure from precision or correctness stemming from a fault.

Definition 2.3 A **failure** entails the absence of the expected or anticipated execution of a particular action.

To illustrate this terminology, consider a memory defect in a computer system due to deterioration, which constitutes a fault. This localized fault can cause errors, such as a bit-flip in a memory cell, leading to a malfunction in a calculation during a user application's read operation, resulting in a failure [87, 107].

Memristive devices are regarded as being in an immature state, affected by noticeable defect densities, manufacturing variations, and susceptibility to temperature and voltage fluctuations. Imperfections stemming from various sources, such as spot defects, assembly discrepancies, and fabrication intricacies, significantly impact not only production yield but also the reliability of memristive devices [3]. Such influence results in parametric and logic faults, which can be classified into two distinct categories: *soft faults* and *hard faults*. Soft faults arise from Cycle-to-Cycle (C2C) or Device-to-Device (D2D) variations, in-field read/write operations, and instances of retention faults where the cell content changes over time. Hard faults may arise from variations during the fabrication process, as well as spot defects, extreme parametric deviations, and continuous stress leading to the device being open or shorted [93].

2.3.2   Defects

The manufacturing of ReRAMs can be divided into two primary phases: the Front End Of Line (FEOL) and the Back End Of Line (BEOL) processes [28]. As depicted in Fig. 2.3a, the FEOL encompasses the creation of transistors, while the BEOL involves the fabrication of metal layers and the ReRAM device. The ReRAM device is typically positioned between metal layers M4 and M5 [33]. During the FEOL phase, transistors are created on the wafer following the standard CMOS process flow. Although this process is well-established, various defects could disrupt the functioning of the transistors, such as patterning proximity effects, line-edge and line-width roughness, polish variations, and

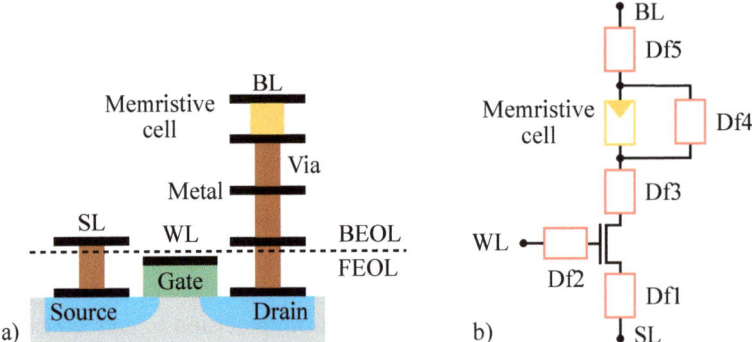

Fig. 2.3 Overview of (**a**) process schematic of an integrated ReRAM cell, and (**b**) fault model of a 1T1R memory cell, utilizing resistors to emulate defects [93]

fluctuations in the gate dielectric [50]. Moving on to the BEOL phase, the fabrication starts with the creation of metal layers. Impurity deposition commonly affects these metal layers, leading to defects at the electrical level [39]. Such imperfections cause resistive open defects in the metal lines connecting the source, drain, and gate of the transistor, as shown in the defect model in Fig. 2.3b. Finally, the ReRAM device is placed between the metal layers. This process involves the deposition of various materials, each of which holds the potential to introduce defects that hinder the proper functioning of the memory cell. The bottom electrode's deposition can be influenced by chemical and physical conditions, significantly impacting the forming process and, consequently, the quality of the LRS. The resistive switching material is also susceptible to various issues, leading to defects, such as thick or thin local spots. Additionally, the top electrode could induce parameter variations and defects [26, 93].

2.3.3 Fault Models

Fault models are a way to abstract physical phenomena for understanding the impact of defects on a system [87]. Several fault models have been introduced to describe the issues related to flawed ReRAMs. Given the similarities between ReRAMs and traditional memories, many traditional fault models are applicable to ReRAMs. These fault models can be grouped into two main categories: *traditional fault models* and *unique fault models* [11, 26, 27].

Traditional Fault Models In the following, fault models are outlined typically observed in both traditional memories (like Static Random-Access Memories (SRAMs) and Dynamic Random-Access Memories (DRAMs)) and ReRAMs.

- **Stuck-at-Fault (SAF)**: The cell consistently resides either in the LRS, designated as *Stuck-at-0*, or in the HRS, termed as *Stuck-at-1*. A Stuck-at-0 condition, classified as a hard fault, materializes when these faults stem from factors like the existence of substantial resistive open defects along the WL (Df2), a persistent open switch (characterized by the access transistor being predominantly in the OFF state), or the presence of a sufficiently large resistive defect, denoted as Df4, which diverts the current in the resistive device (see Fig. 2.3b). These SAFs can also manifest as soft faults if induced by an erroneous forming process. Over-forming the cell results in a Stuck-at-0, signifying that the LRS value falls below its nominal threshold, potentially leading to an incomplete RESET operation due to limitations in the write driver's strength. In the event of a complete failure during the forming operation, resistive switching remains inert, rendering the cell in a Stuck-at-1 [11, 35, 93].
- **Read Destructive Fault/Deceptive Read Destructive Fault (RDF/DRDF)**: During a read operation, the data held within the cell changes as a consequence of the read process and the cell returns the incorrect output. In comparison, the DRDF returns a correct output, but also changes the data held within the cell. Such anomalies predominantly constitute soft faults and manifest in cells with diminished strength, indicated by LRS (HRS) values surpassing (falling below) the designated norm [12, 19, 93].
- **Incorrect Read Faults (IRF)**: The cell returns an incorrect output, although the data stored within the cell remains correct and unaffected by the read process. These irregularities predominantly constitute hard faults, attributed to resistive defects within the memory cell, exemplified by Df3, Df4, and Df5 in Fig. 2.3b [36, 93].
- **Slow Write Fault (SWF)**: Slow write faults are characterized by the inability to successfully complete the write operation within the designated time. These faults can be classified as hard faults if they arise from minor resistive defects at Df2, Df3, and Df4 in Fig. 2.3b. Alternatively, they may be categorized as soft faults when they stem from factors such as a weak access transistor, inadequate capping layer deposition, incorrect stack etching, or due to aging [74, 93].
- **Coupling Fault (CF)**: Coupling faults occur when a write/read operation on one memory cell inadvertently triggers a write operation in an adjacent cell. These faults are categorized as either soft or hard, depending on their origin. Factors such as electromagnetic interference, crosstalk, or manufacturing imperfections can create unintended electrical pathways or influence between adjacent cells, leading to these coupling faults. [10, 58].

Unique Fault Models Resistive Random-Access Memories have distinct malfunction patterns that require unique fault models to describe their behavior.

- **Undefined Write Fault (UWF)**: During a writing operation, the cell enters an undefined state, positioned between the states LRS and HRS. This anomaly is triggered by an insufficient bias voltage during the writing process, particularly occurring in cells with weaker characteristics, possibly magnified by C2C variability. Consequently, if a read operation is conducted on this particular cell, an arbitrary logic value will be obtained [35]. This fault's manifestation is twofold: it can be detected in a newly manufactured cell, attributed to pronounced process variations such as inadequate forming, or in a cell that has aged, resulting from the gradual shift in resistance over time [27, 75, 93].

- **Deep State Fault (Deep):** The cell's resistance exceeds its designated limits, manifesting as a scenario where the resistance in the LRS falls below R_{LRS}, while it exceeds the set limit in the HRS [46]. Such a vulnerability may be attributed to factors like over-forming or variations in C2C characteristics [27].

- **Unknown Read Fault (URF):** When a read operation is conducted, the output yields an arbitrary logic value, regardless of the conditions of the reading process. This form of vulnerability emerges as a soft fault when the LRS and HRS are situated in proximity to each other and consequently close to the reference resistance [27, 46, 93].

2.3.4 Fault Injection

Fault injection techniques have long been acknowledged as indispensable tools for validating system dependability by analyzing device behavior in the event of a fault occurrence. This subsection outlines fault injection methodologies that have been developed to assess system robustness and behavior under diverse fault scenarios [107]:

- **Hardware-based Fault Injection:** This category involves directly disturbing the hardware at a physical level. This technique encompass varying hardware parameters based on the environment, including heavy ion radiation, electromagnetic interferences, and power supply disturbances [4, 23, 56, 107].

- **Software-based Fault Injection:** With the aim of replicating errors that would arise in hardware due to faults, software-based fault injection operates at a software level. The technique simulates the errors in software that would mimic hardware behavior when affected by faults [4, 23, 56, 107].

- **Simulation-based Fault Injection:** This technique injects faults into high-level simulation models. It enables early assessment of system dependability, especially when only a model of the system exists [49, 107].

- **Emulation-based Fault Injection:** Presenting an alternative to time-intensive simulation-based fault injection, this approach employs Field-Programmable Gate Arrays (FPGAs) for accelerating fault simulation and effective circuit emulation. By utilizing FPGAs, designers can study circuit behavior within the real application environment while considering real-time interactions [23, 68, 107].

- **Hybrid Fault Injection:** This technique combines software-implemented fault injection with hardware monitoring [107].

2.4   Hardware Security

Modern computing hardware is a diverse spectrum of components sourced from various vendors with differing levels of trust. Operating within a mixed-trust environment, these components serve diverse security levels. This complex ecosystem, coupled with the extensive connectivity in modern systems, renders critical hardware resources vulnerable to security threats. To counteract these security threats, hardware security measures are crucially needed [38, 82].

These threats span the entire semiconductor life cycle, from design to recycling, emerging from unintentional design flaws [48, 64], malicious modifications [24, 66, 103], and system side effects [16, 86, 88]. Hardware security threats encompass covert channels [9], side channels [69], hardware Trojans [101], and fault injection attacks [31]. These vulnerabilities target a spectrum of critical components, including cryptographic functions, secure architectures, intellectual property, and machine learning models [38].

To design secure system, hardware security properties encompass formal specifications defining invariant security-related properties. These properties guide security verification tools, limit desirable security attributes and assisting in formulating security countermeasures [80]. Subsequent, the three fundamental hardware security properties are outlined referred to as the CIA triad (confidentiality, integrity, and availability).

1. **Confidentiality** mandates that secret information remains undisclosed when observing public outputs or memory locations. Leaks of sensitive data can occur through system side channels, backdoors, covert channels, or hardware Trojans [38].
2. **Integrity** ensures that trusted data remains unaltered by untrusted entities. Attacks targeting critical memory locations, such as cryptographic keys, program counters, or privilege registers, compromise integrity and serve as stepping stones for subsequent malicious activities [38].
3. **Availability** characterizes a system's capacity to consistently execute its designated operations. This hardware security attribute holds paramount importance, as the absence of availability undermines the assurance of the other two properties. Specifically, Denial-of-Service (DoS) attacks focus on this property, aiming to disable or isolate vital components within the system[78].

The remainder of the chapter introduces three common types of hardware security threats: hardware Trojans, side-channel attacks, and fault injection attacks. Each subsection gives a short introduction of the attack by depicting the working principle and briefly discusses the attack surface.

2.4.1 Hardware Trojans

In contrast to software Trojans, the hardware counterpart poses a significant challenge in their removal, making them a serious and persistent threat to computer systems [43]. Hardware Trojans refer to unauthorized changes made to Integrated Circuits (ICs) by adversaries, introducing undesired functionalities. These alterations exploit the global nature of semiconductor design and manufacturing, causing concerns across multiple sectors such as military, finance, and transportation [90]. The production of ICs involves design, fabrication, and testing. To reduce costs and speed up time-to-market, the fabrication is often outsourced to a foundry, while the design takes advantage of third-party Intellectual Property (IP). As a result, the IC supply chain is vulnerable to various hardware security attacks due to the involvement of potentially malicious third parties [100].

Ensuring the authenticity of chips requires expensive end-to-end trust mechanisms or post-manufacturing validation. Hardware Trojans can impact various ICs, including Application-Specific Integrated Circuits (ASICs), microprocessors, and digital signal processors [90]. Reports from reputable sources like the U.S. Administration, the U.S. Senate, and IEEE Spectrum emphasize the severity of this issue [1,17,91]. Efforts to combat Trojan attacks have focused on three primary solutions: (1) Trojan detection methods, (2) Design For Security (DFS) strategies, and (3) runtime monitoring approaches. Trojan detection methods primarily aim to identify Trojans at the IP level using pre-silicon techniques or nondestructive methods during post-silicon manufacturing tests. DFS methods aim to complicate the insertion of hard-to-detect Trojans or assist in their identification during post-silicon validation. However, Trojan detection and DFS methods often lack complete assurance. In contrast, runtime validation methods involve continuous online monitoring of circuit operation, serving as a final defense against Trojan attacks, aiming to mitigate the impact of activated Trojans [6].

2.4.2 Side-Channel Attacks

Another significant class of hardware attacks involves side-channel attacks, exploiting implementation-specific characteristics to extract secret parameters. These attacks capitalize on unintended physical information leaks, aiming to deduce valuable insights about the operational behavior of the target system [45]. Side-channel attacks work by inferring internal computations through the analysis of external parameters like processing time, power consumption, heat dissipation, and electromagnetic emissions. These attacks pose a particular threat to cryptographic implementations, seeking to expose confidential data such as encryption keys [84]. Meltdown [63] is a well-known example of a hardware security breach that exploits a side-channel attack. This attack enables an attacker to gain unauthorized access to the memory of other processes.

Addressing side-channel attacks does not have a general solution to safeguard a computing system. However, specific design alterations can mitigate distinct information leakage and serve as countermeasures against particular side-channel attacks. For instance, the randomization of operation-dependent values is a typical countermeasure to prevent electromagnetic and power side-channel attacks [84]. Additionally, timing side-channel vulnerabilities can be mitigated by equalizing response times, potentially achieved by delaying operations.

2.4.3   Fault Injection Attacks

Fault injection attacks are considered active physical assaults aiming to maliciously extract cryptographic keys, elevate privileges, or compromise the implementation of neural networks. Essentially, an adversary deliberately injects faults into a computer system and observes the system's response to extract sensitive information. Several fault injection methods have been proven to be effective, including clock/voltage glitching and optical/electromagnetic disturbances. Of particular note, the Rowhammer attack [47] has garnered significant attention due to its impact on DRAMs. This attack allows an adversary to intentionally manipulate bits in nearby memory regions by inducing disruptive errors in modern high-density memories.

DRAM cells consist of a capacitor linked to a transistor, with the capacitor's charge encoding two distinct states (refer to Fig. 2.4). While all the transistors' gates within a specific row are interconnected via a WL, the corresponding capacitors of a column are linked through a BL. Unfortunately, the charge of these capacitors is typically transient, limiting the retention time. Consequently, the memory controller must continuously refresh the charge of all memory cells to maintain the stored information's integrity. The

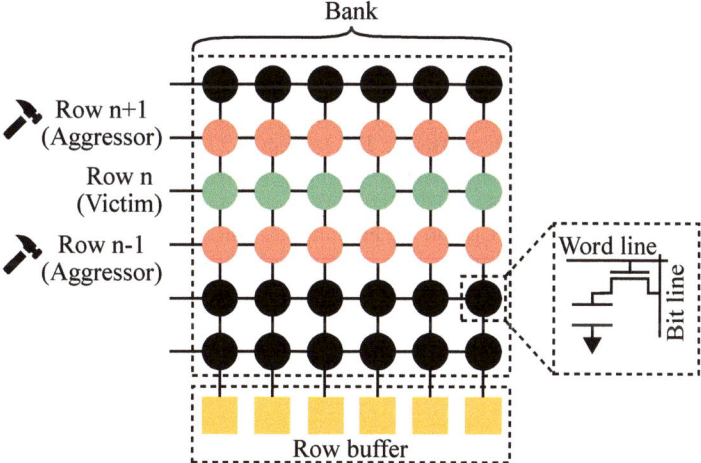

Fig. 2.4 Overview of the Rowhammer attack procedure in DRAMs: Hammering the two adjacent rows surrounding the victim row to intentionally trigger bit-flip faults by deliberately diminishing the capacitor's charge

Rowhammer attack exploits this phenomenon to purposefully diminish the charge of the targeted cell. The disruptive error arises from repetitively targeting a WL, increasing the discharge of neighboring cells, as depicted in Fig. 2.4. Rowhammer has been used to gain kernel privileges, enabling Google Project Zero researchers to effectively take control of entire computer systems [81].

2.5   Synopsis

This chapter lays the groundwork for understanding the book by discussing eNVMs, CIM paradigms, memristor reliability, and hardware security. It starts by explaining memristive devices, which are central to eNVMs, covering their operation, modeling, and applications in high-density memory structures like crossbar arrays. It then transitions to CIM, detailing how memristive crossbar arrays enable analog and logic operations within memory, presenting a more efficient alternative to traditional computing architectures. The reliability of memristive devices, particularly ReRAMs, is outlined next. The chapter discusses the various faults that can occur in ReRAMs due to manufacturing imperfections or operational stresses, and it introduces fault injection as a method to evaluate system robustness in the face of such faults. Finally, the chapter addresses the critical issue of hardware security. It covers the spectrum of threats across the hardware lifecycle, from design and manufacturing to deployment, includes hardware Trojans, side-channel attacks, and fault injection attacks. The chapter concludes by emphasizing the importance of confidentiality, integrity, and availability (the CIA triad) in designing secure systems and the ongoing challenges in protecting against sophisticated hardware attacks.

References

1. Adee, S.: The hunt for the kill switch. IEEE Spectrum **45**(5), 34–39 (2008). https://doi.org/10. 1109/mspec.2008.4505310
2. Ali, K.A., Rizk, M., Baghdadi, A., Diguet, J.P., Jomaah, J., Onizawa, N., Hanyu, T.: Memristive computational memory using memristor overwrite logic (MOL). IEEE Trans. Very Large Scale Integr. Syst. **28**(11), 2370–2382 (2020). https://doi.org/10.1109/tvlsi.2020.3011522
3. Anghel, L., Bernasconi, A., Ciriani, V., Frontini, L., Trucco, G., Vatajelu, I.: Stuck-at fault mitigation of emerging technologies based switching lattices. J. Electron. Testing **36**(3), 313–326 (2020). https://doi.org/10.1007/s10836-020-05885-2
4. Arlat, J., Crouzet, Y., Karlsson, J., Folkesson, P., Fuchs, E., Leber, G.: Comparison of physical and software-implemented fault injection techniques. IEEE Trans. Comput. **52**(9), 1115–1133 (2003). https://doi.org/10.1109/tc.2003.1228509
5. Bhattacharjee, D., Dutt, A., Chattopadhyay, A.: MAMI: majority and multi-input logic on memristive crossbar array. In: IEEE Asia Pacific Conference on Circuits AND Systems (APCCAS). IEEE (2018). https://doi.org/10.1109/apccas.2018.8605573
6. Bhunia, S., Hsiao, M.S., Banga, M., Narasimhan, S.: Hardware trojan attacks: threat analysis and countermeasures. Proc. IEEE **102**(8), 1229–1247 (2014). https://doi.org/10.1109/jproc. 2014.2334493

7. Borghetti, J., Snider, G.S., Kuekes, P.J., Yang, J.J., Stewart, D.R., Williams, R.S.: Memristive switches enable stateful logic operations via material implication. Nature **464**(7290), 873–876 (2010). https://doi.org/10.1038/nature08940

8. Burr, G.W., Shelby, R.M., Sebastian, A., Kim, S., Kim, S., Sidler, S., Virwani, K., Ishii, M., Narayanan, P., Fumarola, A., Sanches, L.L., Boybat, I., Gallo, M.L., Moon, K., Woo, J., Hwang, H., Leblebici, Y.: Neuromorphic computing using non-volatile memory. Adv. Phys. X **2**(1), 89–124 (2016)

9. Caviglione, L.: Trends and challenges in network covert channels countermeasures. Appl. Sci. **11**(4), 1641 (2021). https://doi.org/10.3390/app11041641

10. Chang, M.F., Fuchs, W., Patel, J.: Diagnosis and repair of memory with coupling faults. IEEE Trans. Comput. **38**(4), 493–500 (1989). https://doi.org/10.1109/12.21142

11. Chen, Y.X., Li, J.F.: Fault modeling and testing of 1T1R memristor memories. In: IEEE VLSI Test Symposium (VTS). IEEE (2015). https://doi.org/10.1109/vts.2015.7116247

12. Chen, C.Y., Shih, H.C., Wu, C.W., Lin, C.H., Chiu, P.F., Sheu, S.S., Chen, F.T.: RRAM defect modeling and failure analysis based on march test and a novel squeeze-search scheme. IEEE Trans. Comput. **64**(1), 180–190 (2015). https://doi.org/10.1109/tc.2014.12

13. Chen, Z., Wu, H., Gao, B., Wu, D., Deng, N., Qian, H., Lu, Z., Haukness, B., Kellam, M., Bronner, G.: Performance improvements by SL-current limiter and novel programming methods on 16MB RRAM chip. In: International Memory Workshop (IMW). IEEE (2017). https://doi.org/10.1109/imw.2017.7939097

14. Chou, C., Lin, Z., Tseng, P., Li, C., Chang, C., Chen, W., Chih, Y., Chang, T.J.: An N40 256k×44 embedded RRAM macro with Sl-precharge SA and low-voltage current limiter to improve read and write performance. In: IEEE International Solid - State Circuits Conference - (ISSCC), pp. 478–480. IEEE International Solid - State Circuits Conference - (ISSCC) (2018)

15. Chou, C.C., Lin, Z.J., Lai, C.A., Su, C.I., Tseng, P.L., Chen, W.C., Tsai, W.C., Chu, W.T., Ong, T.C., Chuang, H., Chih, Y.D., Chang, T.Y.J.: A 22nm 96kx144 RRam macro with a self-tracking reference and a low ripple charge pump to achieve a configurable read window and a wide operating voltage range. In: Symposium on VLSI Circuits. IEEE (2020). https://doi.org/10.1109/vlsicircuits18222.2020.9163014

16. Choudary, O., Kuhn, M.G.: Template attacks on different devices. In: Constructive Side-Channel Analysis and Secure Design, pp. 179–198. Springer International Publishing, Berlin (2014). https://doi.org/10.1007/978-3-319-10175-0_13

17. Congressional Record (Senate): The national security aspects of the global migration of the U.S. semiconductor industry (2003). https://irp.fas.org/congress/2003_cr/s060503.html. Accessed Aug 24, 2023

18. Contreras, J.R., Schubert, J., Kohlstedt, H., Waser, R.: Memory device based on a ferroelectric tunnel junction. In: Device Research Conference, Santa Barbara, CA, USA, 24/06/2002-26/06/2002, pp. 97–8. IEEE, Inst fur Festkorperforschung, Forschungszentrum Julich GmbH, Germany, Piscataway, NJ, USA (2002)

19. Dilillo, L., Girard, P., Pravossoudovitch, S., Virazel, A., Borri, S., Hage-Hassan, M.: Dynamic read destructive fault in embedded-SRAMs: analysis and march test solution. In: IEEE European Test Symposium (ETS). IEEE (2004). https://doi.org/10.1109/etsym.2004.1347645

20. Esatu, T.K., Prakash, A., Li, Z., Lau, D., Jo, S.H., Liu, T.J.K.: Highly reliable and secure PUF using resistive memory integrated into a 28 nm CMOS process. Trans. Electron Devices **70**(5), 2291–2296 (2023). https://doi.org/10.1109/ted.2023.3251953

21. Escudero, M., Vourkas, I., Rubio, A., Moll, F.: Variability-tolerant memristor-based ratioed logic in crossbar array. In: International Symposium on Nanoscale Architectures, NANOARCH '18. ACM (2018). https://doi.org/10.1145/3232195.3232213

22. Escudero, M., Vourkas, I., Rubio, A., Moll, F.: Memristive logic in crossbar memory arrays: variability-aware design for higher reliability. IEEE Trans. Nanotechnol. **18**, 635–646 (2019). https://doi.org/10.1109/tnano.2019.2923731

23. Eslami, M., Ghavami, B., Raji, M., Mahani, A.: A survey on fault injection methods of digital integrated circuits. Integration **71**, 154–163 (2020). https://doi.org/10.1016/j.vlsi.2019.11.006

24. Fern, N., Kulkarni, S., Cheng, K.T.T.: Hardware trojans hidden in RTL don't cares - automated insertion and prevention methodologies. In: IEEE International Test Conference (ITC). IEEE (2015). https://doi.org/10.1109/test.2015.7342387

25. FernANDez, C., Vourkas, I.: Reliability-aware ratioed logic operations for energy-efficient computational ReRAM. In: International Conference on Very Large Scale Integration (VLSI-SoC). IEEE (2022). https://doi.org/10.1109/vlsi-soc54400.2022.9939627

26. Fieback, M.: Testing RRAM and computation-in-memory devices. Ph.D. Thesis (2022). https://doi.org/10.4233/UUID:71E1CBF8-CE02-4CE8-A09B-883E6F84E996

27. Fieback, M., Taouil, M., Hamdioui, S.: Testing resistive memories: where are we and what is missing? In: IEEE International Test Conference (ITC). IEEE (2018). https://doi.org/10.1109/test.2018.8624895

28. Fieback, M., Medeiros, G.C., Wu, L., Aziza, H., Bishnoi, R., Taouil, M., Hamdioui, S.: Defects, fault modeling, and test development framework for RRAMs. ACM J. Emerg. Technol. Comput. Syst. **18**(3), 1–26 (2022). https://doi.org/10.1145/3510851

29. Funck, C., Menzel, S.: Comprehensive model of electron conduction in oxide-based memristive devices. ACS Appl. Electron. Mater. **3**, 3674–3692 (2021)

30. Gallo, M.L., Sebastian, A.: An overview of phase-change memory device physics. J. Phys. D Appl. Phys. **53**(21), 213002 (2020)

31. Gangolli, A., Mahmoud, Q.H., Azim, A.: A systematic review of fault injection attacks on IoT systems. Electronics **11**(13), 2023 (2022). https://doi.org/10.3390/electronics11132023

32. Grenouillet, L., Castellani, N., Persico, A., Meli, V., Martin, S., Billoint, O., Segaud, R., Bernasconi, S., Pellissier, C., Jahan, C., Charpin-Nicolle, C., Dezest, P., Carabasse, C., Besombes, P., Ricavy, S., Tran, N.P., Magalhaes-Lucas, A., Roman, A., Boixaderas, C., Magis, T., Bedjaoui, M., Tessaire, M., Seignard, A., Mazen, F., LANDis, S., Vianello, E., Molas, G., Gaillard, F., Arcamone, J., Nowak, E.: 16kbit 1T1R oxram arrays embedded in 28nm fdsoi technology demonstrating low ber, high endurance, and compatibility with core logic transistors. In: International Memory Workshop (IMW). IEEE (2021). https://doi.org/10.1109/imw51353.2021.9439607

33. Grossi, A., Nowak, E., Zambelli, C., Pellissier, C., Bernasconi, S., Cibrario, G., Hajjam, K.E., Crochemore, R., Nodin, J., Olivo, P., Perniola, L.: Fundamental variability limits of filament-based RRAM. In: IEEE International Electron Devices Meeting (IEDM). IEEE (2016). https://doi.org/10.1109/iedm.2016.7838348

34. Guckert, L., SwartzlANDer, E.E.: MAD gates - memristor logic design using driver circuitry. IEEE Trans. Circuits Syst. II Express Briefs **64**(2), 171–175 (2017). https://doi.org/10.1109/tcsii.2016.2551554

35. Haron, N.Z., Hamdioui, S.: On defect oriented testing for hybrid CMOS/memristor memory. In: Asian Test Symposium. IEEE (2011). https://doi.org/10.1109/ats.2011.66

36. Haron, N.Z., Hamdioui, S.: DfT schemes for resistive open defects in RRAMs. In: Design, Automation & Test in Europe Conference & Exhibition (DATE). IEEE (2012). https://doi.org/10.1109/date.2012.6176603

37. Hoffer, B., Wainstein, N., Neumann, C.M., Pop, E., Yalon, E., Kvatinsky, S.: Stateful logic using phase change memory. IEEE J. Exploratory Solid-State Comput. Devices Circuits **8**(2), 77–83 (2022). https://doi.org/10.1109/jxcdc.2022.3219731

38. Hu, W., Chang, C.H., Sengupta, A., Bhunia, S., Kastner, R., Li, H.: An overview of hardware security and trust: threats, countermeasures, and design tools. IEEE Trans. Comput.-Aided Design Integr. Circuits Syst. **40**(6), 1010–1038 (2021). https://doi.org/10.1109/tcad.2020. 3047976

39. Ielmini, D., Waser, R.: Resistive Switching: From Fundamentals of Nanoionic Redox Processes to Memristive Device Applications. Wiley-VCH Verlag GmbH & Co. KGaA, Weinheim (2016). https://doi.org/10.1002/9783527680870

40. Ielmini, D., Wong, H.P.: In-memory computing with resistive switching devices. Nature Electron. **1**(6), 333–343 (2018)

41. Intel Corporation: Intel Optane Memory Series (2024). https://www.intel.com/content/www/ us/en/products/sku/97544/intel-optane-memory-series-16gb-m-2-80mm-pcie-3-0-20nm-3d-xpoint/specifications.html. Accessed: May 22, 2023

42. Jain, P., Arslan, U., Sekhar, M., Lin, B.C., Wei, L., Sahu, T., Alzate-vinasco, J., Vangapaty, A., Meterelliyoz, M., Strutt, N., Chen, A.B., Hentges, P., Quintero, P.A., Connor, C., Golonzka, O., Fischer, K., Hamzaoglu, F.: 13.2 a 3.6Mb 10.1Mb/mm2 embedded non-volatile ReRAM macro in 22nm FinFET technology with adaptive forming/set/reset schemes yielding down to 0.5V with sensing time of 5ns at 0.7V. In: International Solid- State Circuits Conference (ISSCC). IEEE (2019). https://doi.org/10.1109/isscc.2019.8662393

43. Jin, Y.: Introduction to hardware security. Electronics **4**(4), 763–784 (2015). https://doi.org/10. 3390/electronics4040763

44. Jo, S.H., Kumar, T., Zitlaw, C., Nazarian, H.: Self-limited rram with on/off resistance ratio amplification. In: Symposium on VLSI Technology (VLSI Technology). IEEE (2015). https:// doi.org/10.1109/vlsit.2015.7223715

45. Joy Persial, G., Prabhu, M., Shanmugalakshmi, R.: Side channel attack-survey. Int. J. Sci. Res. Rev. **1**(4), 54–57 (2011)

46. Kannan, S., Rajendran, J., Karri, R., Sinanoglu, O.: Sneak-path testing of memristor-based memories. In: International Conference on VLSI Design AND International Conference on Embedded Systems. IEEE (2013). https://doi.org/10.1109/vlsid.2013.219

47. Kim, Y., Daly, R., Kim, J., Fallin, C., Lee, J.H., Lee, D., Wilkerson, C., Lai, K., Mutlu, O.: Flipping bits in memory without accessing them. ACM SIGARCH Comput. Archit. News **42**(3), 361–372 (2014). https://doi.org/10.1145/2678373.2665726

48. Kocher, P., Horn, J., Fogh, A., Genkin, D., Gruss, D., Haas, W., Hamburg, M., Lipp, M., Mangard, S., Prescher, T., Schwarz, M., Yarom, Y.: Spectre attacks: exploiting speculative execution. In: IEEE Symposium on Security AND Privacy (S&P). IEEE (2019). https://doi. org/10.1109/sp.2019.00002

49. Kooli, M., Natale, G.D.: A survey on simulation-based fault injection tools for complex systems. In: IEEE International Conference on Design & Technology of Integrated Systems in Nanoscale Era (DTIS). IEEE (2014). https://doi.org/10.1109/dtis.2014.6850649

50. Kuhn, K.J., Giles, M.D., Becher, D., Kolar, P., Kornfeld, A., Kotlyar, R., Ma, S.T., Maheshwari, A., Mudanai, S.: Process technology variation. IEEE Trans. Electron Devices **58**(8), 2197–2208 (2011). https://doi.org/10.1109/ted.2011.2121913

51. Kvatinsky, S., Kolodny, A., Weiser, U.C., Friedman, E.G.: Memristor-based IMPLY logic design procedure. In: IEEE International Conference on Computer Design (ICCD). IEEE (2011). https://doi.org/10.1109/iccd.2011.6081389

52. Kvatinsky, S., Wald, N., Satat, G., Kolodny, A., Weiser, U.C., Friedman, E.G.: MRL - memristor ratioed logic. In: International Workshop on Cellular Nanoscale Networks and Their Applications. IEEE (2012). https://doi.org/10.1109/cnna.2012.6331426

53. Kvatinsky, S., Belousov, D., Liman, S., Satat, G., Wald, N., Friedman, E.G., Kolodny, A., Weiser, U.C.: MAGIC - memristor-aided logic. IEEE Trans. Circuits Syst. II Express Briefs **61**(11), 895–899 (2014). https://doi.org/10.1109/tcsii.2014.2357292

54. Le, B.Q., Levy, A., Wu, T.F., Radway, R.M., Hsieh, E.R., Zheng, X., Nelson, M., Raina, P., Wong, H.S.P., Wong, S., Mitra, S.: Radar: a fast and energy-efficient programming technique for multiple bits-per-cell Rram arrays. IEEE Trans. Electron Devices **68**(9), 4397–4403 (2021)
55. Lehtonen, E., Laiho, M.: Stateful implication logic with memristors. In: IEEE/ACM International Symposium on Nanoscale Architectures. IEEE (2009). https://doi.org/10.1109/nanoarch.2009.5226356
56. Lenka, R.K., Padhi, S., Nayak, K.M.: Fault injection techniques - a brief review. In: International Conference on Advances in Computing, Communication Control AND Networking (ICACCCN). IEEE (2018). https://doi.org/10.1109/icacccn.2018.8748585
57. Levy, Y., Bruck, J., Cassuto, Y., Friedman, E.G., Kolodny, A., Yaakobi, E., Kvatinsky, S.: Logic operations in memory using a memristive akers array. Microelectron. J. **45**(11), 1429–1437 (2014). https://doi.org/10.1016/j.mejo.2014.06.006
58. Li, J.F., Cheng, K.L., Huang, C.T., Wu, C.W.: March-based RAM diagnosis algorithms for stuck-at and coupling faults. In: Proceedings International Test Conference. IEEE (2001). https://doi.org/10.1109/test.2001.966697
59. Li, C., Li, Y., Jiang, H., Song, W., Lin, P., Wang, Z., Yang, J.J., Xia, Q., Hu, M., Montgomery, E., Zhang, J., Davila, N., Graves, C.E., Li, Z., Strachan, J.P., Williams, R.S., Ge, N., Barnell, M., Wu, Q.: Large memristor crossbars for analog computing. In: International Symposium on Circuits and Systems (ISCAS). IEEE (2018). https://doi.org/10.1109/iscas.2018.8351877
60. Li, H., Chen, W., Levy, A., Wang, C., Wang, H., Chen, P., Wan, W., Khwa, W., Chuang, H., Chih, Y., Chang, M., Wong, H.P., Raina, P.: Sapiens: A 64-Kb RRAM non-volatile associative memory for one-shot learning and inference at the edge. IEEE Trans. Electron Devices 1–7 (2021)
61. Liaw, C.: Integrated semiconductor memory with an arrangement of nonvolatile memory cells, and method. United States Patent 7277312 (2007)
62. Linn, E., Rosezin, R., Kügeler, C., Waser, R.: Complementary resistive switches for passive nanocrossbar memories. Nature Mater. **9**(5), 403–406 (2010)
63. Lipp, M., Schwarz, M., Gruss, D., Prescher, T., Haas, W., Fogh, A., Horn, J., Mangard, S., Kocher, P., Genkin, D., Yarom, Y., Hamburg, M.: Meltdown: reading kernel memory from user space. In: USENIX Security Symposium (USENIX Security 18), pp. 973–990. USENIX Association, Baltimore, MD (2018). https://www.usenix.org/conference/usenixsecurity18/presentation/lipp
64. Lipp, M., Schwarz, M., Gruss, D., Prescher, T., Haas, W., Horn, J., Mangard, S., Kocher, P., Genkin, D., Yarom, Y., Hamburg, M., Strackx, R.: Meltdown: reading kernel memory from user space. Commun. ACM **63**(6), 46–56 (2020). https://doi.org/10.1145/3357033
65. Liu, C., Hu, M., Strachan, J.P., Li, H.H.: Rescuing memristor-based neuromorphic design with high defects. In: Proceedings of the Design Automation Conference. ACM (2017). https://doi.org/10.1145/3061639.3062310
66. Liu, Y., Jin, Y., Nosratinia, A., Makris, Y.: Silicon demonstration of hardware trojan design and detection in wireless cryptographic ICs. IEEE Trans. Very Large Scale Integr. Syst. **25**(4), 1506–1519 (2017). https://doi.org/10.1109/tvlsi.2016.2633348
67. Liu, M., Chakrabarty, K.: Online fault detection in ReRAM-based computing systems for inferencing. IEEE Trans. Very Large Scale Integr. Syst. **30**(4), 392–405 (2022). https://doi.org/10.1109/tvlsi.2021.3139530
68. Lopez-Ongil, C., Entrena, L., Garcia-Valderas, M., Portela, M., Aguirre, M., Tombs, J., Baena, V., Munoz, F.: A unified environment for fault injection at any design level based on emulation. IEEE Trans. Nucl. Sci. **54**(4), 946–950 (2007). https://doi.org/10.1109/tns.2007.904078
69. Lyu, Y., Mishra, P.: A survey of side-channel attacks on caches and countermeasures. J. Hardware Syst. Security **2**(1), 33–50 (2017). https://doi.org/10.1007/s41635-017-0025-y

70. Marchewka, A., Roesgen, B., Skaja, K., Du, H., Jia, C.L., Mayer, J., Rana, V., Waser, R., Menzel, S.: Nanoionic resistive switching memories: on the physical nature of the dynamic reset process. ACS Appl. Electron. Mater. **2**(1), 1500233/1–13 (2016)

71. Menzel, S., Waters, M., Marchewka, A., Böttger, U., Dittmann, R., Waser, R.: Origin of the ultra-nonlinear switching kinetics in oxide-based resistive switches. Adv. Funct. Mater. **21**(23), 4487–4492 (2011)

72. Menzel, S., von Witzleben, M., Havel, V., Boettger, U.: The ultimate switching speed limit of redox-based restive switching devices. Faraday Discuss. **213**, 197–213 (2019)

73. Motaman, S., Ghosh, S.: Dynamic computing in memory (DCIM) in resistive crossbar arrays. In: IEEE International Conference on Computer Design (ICCD). IEEE (2018). https://doi.org/10.1109/iccd.2018.00036

74. Mozaffari, S.N., Tragoudas, S., Haniotakis, T.: Fast march tests for defects in resistive memory. In: IEEE/ACM International Symposium on Nanoscale Architectures (NANOARCH). IEEE (2015). https://doi.org/10.1109/nanoarch.2015.7180592

75. Nguyen, C., Cagli, C., Molas, G., Sklenard, B., Nail, C., Hajjam, K.E., Nodin, J.F., Charpin, C., Bernasconi, S., Reimbold, G.: Study of forming impact on 4Kbit RRAM array performances and reliability. In: IEEE International Memory Workshop (IMW). IEEE (2017). https://doi.org/10.1109/imw.2017.7939105

76. Otsuka, W., Miyata, K., Kitagawa, M., Tsutsui, K., Tsushima, T., Yoshihara, H., Namise, T., Terao, Y., Ogata, K.: A 4mb conductive-bridge resistive memory with 2.3gb/s read-throughput and 216mb/s program-throughput. In: International Solid-State Circuits Conference. IEEE (2011). https://doi.org/10.1109/isscc.2011.5746286

77. Parkin, S., Jiang, X., Kaiser, C., Panchula, A., Roche, K., Samant, M.: Magnetically engineered spintronic sensors and memory. Proc. IEEE **91**(5), 661–680 (2003)

78. Qadir, S., Quadri, S.M.K.: Information availability: an insight into the most important attribute of information security. J. Inf. Security **07**(03), 185–194 (2016). https://doi.org/10.4236/jis.2016.73014

79. Reuben, J., Ben-Hur, R., Wald, N., Talati, N., Ali, A.H., Gaillardon, P.E., Kvatinsky, S.: Memristive logic: a framework for evaluation and comparison. In: International Symposium on Power AND Timing Modeling, Optimization AND Simulation (PATMOS). IEEE (2017). https://doi.org/10.1109/patmos.2017.8106959

80. Samonas, S., Coss, D.: The CIA strikes back: redefining confidentiality, integrity and availability in security. J. Inf. Syst. Security **10**(3) (2014)

81. Seaborn, M., Dullien, T.: Exploiting the DRAM rowhammer bug to gain Kernel privileges (2015). http://googleprojectzero.blogspot.com.tr/2015/03/exploiting-dram-rowhammer-bug-to-gain.html

82. Sisejkovic, D., Leupers, R.: Logic locking: a practical approach to secure hardware. Springer International Publishing, Berlin (2023). https://doi.org/10.1007/978-3-031-19123-7

83. Son, S., La Torre, C., Kindsmüller, A., Rana, V., Menzel, S.: A study of the electroforming process in 1T1R memory arrays. IEEE Trans. Comput.-Aided Design Integr. Circuits Syst. **42**(2), 558–568 (2023). https://doi.org/10.1109/TCAD.2022.3175947

84. Spreitzer, R., Moonsamy, V., Korak, T., Mangard, S.: Systematic classification of side-channel attacks: a case study for mobile devices. IEEE Commun. Surv. Tutor. **20**(1), 465–488 (2018). https://doi.org/10.1109/comst.2017.2779824

85. Staudigl, F., Merchant, F., Leupers, R.: A survey of neuromorphic computing-in-memory: architectures, simulators, and security. IEEE Design Test **39**(2), 90–99 (2022). https://doi.org/10.1109/mdat.2021.3102013

86. Szefer, J.: Survey of microarchitectural side and covert channels, attacks, and defenses. J. Hardware Syst. Security **3**(3), 219–234 (2018). https://doi.org/10.1007/s41635-018-0046-1

87. Tahoor, M.: Reliable computing I, lecture 3: faults, errors, failures (2016). https://cdnc.itec.kit. edu/downloads/lecture3-reliable-computing-1-2016-2017.pdf. Accessed Aug 14, 2023

88. Tang, A., Sethumadhavan, S., Stolfo, S.: CLKSCREW: exposing the perils of security-oblivious energy management. In: USENIX security symposium (USENIX security 17), pp. 1057–1074. USENIX Association, Vancouver (2017)

89. Tang, X., Liu, J., Shen, Y., Li, S., Shen, L., Sanyal, A., Ragab, K., Sun, N.: Low-power SAR ADC design: overview and survey of state-of-the-art techniques. IEEE Trans. Circuits Syst. I Regular Papers **69**(6), 2249–2262 (2022). https://doi.org/10.1109/tcsi.2022.3166792

90. Tehranipoor, M., Koushanfar, F.: A survey of hardware trojan taxonomy and detection. IEEE Design Test Comput. **27**(1), 10–25 (2010). https://doi.org/10.1109/mdt.2010.7

91. The White House: CHIPS and science act will lower costs, create jobs, strengthen supply chains, and counter China (2022). https://www.whitehouse.gov/briefing-room/ statements-releases/2022/08/09/fact-sheet-chips-AND-science-act-will-lower-costs-create-jobs-strengthen-supply-chains-AND-counter-china/. Accessed Aug 23, 2023

92. Torrezan, A.C., Strachan, J.P., Medeiros-Ribeiro, G., Williams, R.S.: Sub-nanosecond switching of a tantalum oxide memristor. Nanotechnology **22**, 485203 (2011)

93. Vatajelu, E.I., Prinetto, P., Taouil, M., Hamdioui, S.: Challenges and solutions in emerging memory testing. IEEE Trans. Emerg. Topics Comput. **7**(3), 493–506 (2019). https://doi.org/10. 1109/tetc.2017.2691263

94. Waser, R., Aono, M.: Nanoionics-based resistive switching memories. Nature Mater. **6**(11), 833–840 (2007)

95. Waser, R., Dittmann, R., Staikov, G., Szot, K.: Redox-based resistive switching memories - nanoionic mechanisms, prospects, and challenges. Adv. Mater. **21**(25–26), 2632–2663 (2009)

96. Wen, J., Baroni, A., Perez, E., Uhlmann, M., Fritscher, M., KrishneGowda, K., Ulbricht, M., Wenger, C., Krstic, M.: Towards reliable and energy-efficient RRAM based discrete fourier transform accelerator. In: Design, Automation & Test in Europe Conference & Exhibition (DATE), pp. 1–6. IEEE (2024). https://doi.org/10.23919/date58400.2024.10546709

97. von Witzleben, M., Fleck, K., Funck, C., Baumkötter, B., Zuric, M., Idt, A., Breuer, T., Waser, R., Böttger, U., Menzel, S.: Investigation of the impact of high temperatures on the switching kinetics of redox-based resistive switching cells using a highspeed nanoheater. Adv. Electron. Mater. **3**(12), 1700294 (2017)

98. von Witzleben, M., Hennen, T., Kindsmüller, A., Menzel, S., Waser, R., Böttger, U.: Study of the set switching event of VCM-based memories on a picosecond timescale. J. Appl. Phys. **127**(20), 204501 (2020)

99. Xia, Q., Robinett, W., Cumbie, M.W., Banerjee, N., Cardinali, T.J., Yang, J.J., Wu, W., Li, X., Tong, W.M., Strukov, D.B., Snider, G.S., Medeiros-Ribeiro, G., Williams, R.S.: Memristor-CMOS hybrid integrated circuits for reconfigurable logic. Nano Lett. **9**(10), 3640–3645 (2009). https://doi.org/10.1021/nl901874j

100. Xiao, K., Forte, D., Jin, Y., Karri, R., Bhunia, S., Tehranipoor, M.: Hardware trojans: lessons learned after one decade of research. Trans. Design Automation Electron. Syst. **22**(1), 1–23 (2016). https://doi.org/10.1145/2906147

101. Xue, M., Gu, C., Liu, W., Yu, S., O'Neill, M.: Ten years of hardware trojans: a survey from the attacker's perspective. Comput. Digital Tech. IET **14**(6), 231–246 (2020). https://doi.org/ 10.1049/iet-cdt.2020.0041

102. Yang, J.J., Pickett, M.D., Li, X., Ohlberg, D.A.A., Stewart, D.R., Williams, R.S.: Memristive switching mechanism for metal/oxide/metal nanodevices. Nature Nanotechnology **3**(7), 429–433 (2008)

103. Yang, K., Hicks, M., Dong, Q., Austin, T., Sylvester, D.: A2: analog malicious hardware. In: IEEE Symposium on Security AND Privacy (SP). IEEE (2016). https://doi.org/10.1109/sp. 2016.10

104. Yasuda, S., Ohba, K., Mizuguchi, T., Sei, H., Shimuta, M., Aratani, K., Shiimoto, T., Yamamoto, T., Sone, T., Nonoguchi, S., Okuno, J., Kouchiyama, A., Otsuka, W., Tsutsui, K.: A cross point Cu-ReRAM with a novel ots selector for storage class memory applications. In: Symposium on VLSI Technology. IEEE (2017). https://doi.org/10.23919/vlsit.2017.7998189

105. Yoneda, S., Ito, S., Hayakawa, Y., Wei, Z., Muraoka, S., Yasuhara, R., Kawashima, K., Himeno, A., Mikawa, T.: Newly developed process integration technologies for highly reliable 40 Nm ReRAM. Jpn. J. Appl. Phys. **58**, SBBB06/1–8 (2019)

106. Zhang, Y., Shen, Y., Wang, X., Guo, Y.: A novel design for a memristor-based OR gate. IEEE Trans. Circuits Syst. II Express Briefs **62**(8), 781–785 (2015). https://doi.org/10.1109/tcsii.2015.2435354

107. Ziade, H., Ayoubi, R.A., Velazco, R.: A survey on fault injection techniques. Int. Arab J. Inf. Technol. **1**(2), 171–186 (2004)

Related Work

3

Since Leon Chua's introduction of the memristor in 1971, substantial research efforts have been dedicated to establishing efficient and dependable computing systems rooted in memristor technology [10]. Despite these advancements, memristive devices continue to face persistent reliability challenges. Addressing these issues has become a critical area of research, with efforts aimed at understanding the underlying causes and developing strategies to enhance their stability and longevity. Beyond the need for dependable systems, these reliability flaws also create potential vulnerabilities that could be exploited in hardware security attacks, allowing adversaries to manipulate or extract sensitive information.

This chapter provides a detailed overview of the reliability challenges in memristor-based computing systems and their implications for hardware security. Section 3.1 focuses on the reliability issues specific to memristor-based memory systems. Section 3.2 examines how these reliability weaknesses can be exploited in hardware security attacks and discusses possible defense mechanisms. Section 3.3 covers the development of instrumentation platforms, which are crucial for assessing and improving reliability. Finally, Sect. 3.4 summarizes the key lessons learned from the survey.

3.1 Reliability of Neuromorphic Computing Systems

Resistive RAMs utilize memristive crossbar arrays to achieve both high memory density and energy efficiency. These crossbar structures not only enable the storage of binary and multi-value data but also support MAC operations. In the quest to explore the reliability characteristics of ReRAMs, it is imperative to encompass all operational modes. Consequently, this section comprehensively outlines prior research regarding the reliability

© The Author(s), under exclusive license to Springer Nature Switzerland AG 2026
F. Staudigl, R. Leupers, *Towards Trustworthy Neuromorphic Computing*,
Synthesis Lectures on Engineering, Science, and Technology,
https://doi.org/10.1007/978-3-032-09586-2_3

aspects of ReRAMs, considering their role as conventional memory as well as their involvement in in-memory computing paradigms. Furthermore, this section provides an overview of fault injection platforms and encapsulates the proposed techniques designed to enhance the dependability of neuromorphic systems [43].

3.1.1 Reliability Aspects of ReRAMs

The landscape of publications exploring the reliability aspects of ReRAM spans from fault detection to the development of memory architectures optimized for reliability. Since the memristor is the fundamental building block, a multitude of publications have proposed fault models and testing approaches for both binary [6, 7, 16, 21, 32, 39] and multilevel cells [20, 22].

Alongside concerns at the memristor level, the commonly employed crossbar structures introduce challenges such as IR drop issues and read/write disturbances. Zhang et al. [49] introduced a circuit architecture co-optimization framework to address the inherent shortcomings of crossbar structures. They implement a double-sided write driver to diminish IR drops along the bit lines and tackle write disturbances through a disturbance detection scheme. Furthermore, they partition the crossbar structure into multiple regions for storing cold and hot data in slow and fast regions, respectively, enhancing both latency and reliability. The integration of these enhancements yields a 26.1% performance improvement while reducing energy consumption by 21.6%. This approach also enhances system reliability by countering read/write disturbances and the IR drop problem.

While circuit-level optimizations effectively enhance reliability, they tend to increase silicon area and energy consumption. Consequently, Mao et al. [33, 34] investigated the impact of read/write voltages and pulse lengths on ReRAM reliability and endurance. Conventionally, SET and RESET operations use distinct voltages. However, the authors advocate for the use of a single voltage for both operations, reducing write latency and energy consumption. Furthermore, they argue that retention can be extended by adjusting word line, bit line, and select line voltages rather than solely relying on altering the ON/OFF ratio of the memristor.

Conventional memories leverage Error Correcting Codes (ECCs) for error detection and rectification [17]. As a result, various error correction techniques have been proposed to augment ReRAM reliability. Schechter et al. [41] introduced an Error Correction Pointer (ECP) scheme to address hard errors in ReRAMs by substituting faulty cells with new ones and recording the locations of the faulty cells. In contrast, Xu et al. [48] proposed an error-resilient ReRAM architecture utilizing ECC and ECP to mitigate retention failures and stuck-at-faults. Moreover, Zheng et al. [52] introduced a detection and recovery scheme to alleviate pseudo-hard errors in ReRAMs by utilizing a higher programming voltage. The authors define a pseudo-hard error as a type of hard error that is still recoverable. Lastly, Zhang et al. [51] proposed a microarchitectural design termed EnTiered-crossbar, partitioning each crossbar along the bit lines into two halves. These near and far segments are isolated using an access transistor, resolving the IR drop issue.

3.1.2   Reliability Aspects of Computing-in-Memory Applications

Memristive crossbar structures offer not just traditional memory functions but also the ability to conduct analog MAC operations and execute logic gates. However, the reliability challenges related to the underlying memristive cells can significantly impact applications built upon them. Consequently, numerous research initiatives have been conducted to evaluate the impact of these reliability concerns on various applications.

Among the most common operational modes of memristor-based accelerators is the execution of analog MAC operations (see Sect. 2.2). Feinberg et al. [13] introduce an error correction scheme based on arithmetic codes to enhance reliability. Data is encoded by multiplying it with an integer, allowing error detection and correction through modulus operations and a correction table lookup. Additionally, the authors improve their approach by using data-aware encoding, leveraging the state dependence of errors and prioritizing critical computation segments for overall system accuracy.

In addition to data encoding, matrix transformation has been proposed to address stuck-at faults in ReRAM accelerators. Zhang et al. [50] utilize row flipping, permutation, and value range adjustments to fortify weight matrices against stuck-at faults. The row flipping transformation converts stuck-off (stuck-on) faults into stuck-on (stuck-off) faults, while the permutation transformation maps smaller (larger) weights to memristors stuck-off (stuck-on). The value range transformation diminishes extreme element magnitudes in the matrix, thereby reducing errors introduced by each stuck-at fault. Experimental results indicate that this framework can recover 99% of accuracy loss caused by stuck-at faults, eliminating the need for neural network retraining.

Furthermore, mapping algorithms have been annotated to mitigate stuck-at faults by thoroughly exploring the mapping space. Xia et al. [46] introduce an inner fault-tolerant mapping algorithm capable of addressing multiple faulty columns without hardware overhead. The authors use two cells to encode a single value, enabling the algorithm to adjust, in case of a faulty cell, the other cell to represent the correct value. Even before the mapping, the neural network itself can be optimized to mitigate the non-ideal effects of memristive crossbars. Al-Shaarawt et al. [2] introduce the PRUNIX framework for training and pruning convolutional neural networks tailored for deployment on memristor crossbar-based accelerators. PRUNIX addresses various non-ideal characteristics of memristor crossbars, including weight quantization, state-drift, aging, and stuck-at faults. It incorporates a unique Group Sawtooth Regularization to enhance tolerance to non-idealities and promote sparsity. Additionally, it employs the adaptive pruning algorithm to minimize accuracy loss by considering the sensitivity of different CNN layers to pruning.

Finally, Emara et al. [11] discuss production testing of a XOR gate, using memristors in conjunction with CMOS inverters. The research specifically investigates the two-input

XOR gate using a fault model that includes stuck-at faults for memristors and a fault model for transistors. The analysis reveals that faults within the XOR gate generate analog output voltage values due to the circuit's architecture. Consequently, a specialized 2-bit Flash ADC is employed to achieve comprehensive fault coverage. Notably, the study highlights that four resistive short faults in the XOR gate can only be detected by monitoring the input current, emphasizing the need for exhaustive testing to attain 100% fault coverage.

3.1.3   Simulation Platforms

As the landscape of memristor-based memories advances, a multitude of simulation platforms has surfaced to estimate critical system parameters including computing performance, energy consumption, and area. These platforms additionally simulate device variability, process variations, and faults to bolster the reliability of their forecasts. The comprehensive comparative analysis in Table 3.1 serves as the cornerstone for the following discussion in this section.

The majority of publications explore the influence of device variation and variability on the accuracy of Deep Neural Networks (DNNs). Notably, RxNN [19] introduces a swift and precise simulation framework leveraging a Fast Crossbar Model (FCM) that significantly accelerates the evaluation process. This platform adeptly identifies errors stemming from interconnect and sense parasitic, sneak path currents, and synaptic conductance variations during inference and training. The FCM abstracts non-idealities by generating a non-ideal conductance matrix, employing three consecutive matrix transformations to replicate synaptic device characteristics, interconnect/circuit parameters, and the chip variation profile.

Although most simulators adopt a similar approach [28, 31, 37, 40, 45, 53], PytorX [18] distinguishes itself with its comprehensive methodology for conducting end-to-end training, mapping, and evaluation, while considering a wide spectrum of reliability factors. Implemented in Python and built upon the PyTorch library, PytorX replicates MAC operations on memristive crossbars, encompassing considerations of DACs/ADCs, weight mapping, and weight partitioning on multiple arrays. To emulate the impact of the IR-drop in crossbar arrays, the authors included a dynamic crossbar solver based on the simplified modified nodal analysis. However, this approach is computationally intensive for simulating extensive neural networks, leading the authors to propose a noise injection adaption method capable of statistically approximating the effect of the IR-drop.

NeuroSim [8, 9, 36] comprises a group of instruction-accurate simulators building upon a circuit-level macro model that estimates various metrics of neuromorphic architectures, such as area, latency, dynamic energy, and leakage power. This platform facilitates a direct comparison of conventional SRAM cores, digital eNVM cores, and analog eNVM cores. In essence, NeuroSim furnishes the fundamental components to construct a comprehensive hierarchical architecture, considering not only reliability factors of the

Table 3.1 Overview of simulation platforms used for evaluating the reliability of non-volatile memories [5, 29]

Simulation platform	Prog. language	Inference	Training	Open-source	Supported devices	Non-idealities
GENIEx [5]	Python	✓	✗	✗	Non-volatile memories.	Selector/device parasitics, source/sink/wire resistances
CrossSim [47]	Python	✓	✓	✓	Non-volatile memories.	C2C/D2D variations, programming errors, conductance drift, read noise variability, and ADC precision loss.
NeuroSim [9]/ NeuroSim+ [8]/ NeuroSim+DNN [36]	C++, Python	✓	✓	✓	Non-volatile memories, legacy NAND flash.	C2C/D2D variations, device non-linearities.
SySCIM [42]	C++, SystemC	n/a	n/a	✗	Non-volatile memories.	C2C/D2D variations, device, interconnect parasitics, SAF.
IBM analog hardware acceleration kit [38]	C++, Python	✓	✓	✓	Non-volatile memories.	C2C/D2D variations, ADC/DAC discretization, noise, and device fluctuations.
MNSIM [45]/ MNSIM2.0 [53]	Python	✓	✓	✗	Non-volatile memories.	Non-ideal device factors, interconnect parasitics.
DL-RSIM [31]	Python	✓	✗	✗	Non-volatile memories.	Non-ideal circuit and device properties.
MemTorch [28]	C++, Python	✓	✗	✗	Non-volatile memories, legacy NAND flash.	C2C/D2D variations, device failure.
TxSim [40]	Python	✓	✓	✗	Non-volatile memories, legacy NAND flash.	Interconnect parasitic, sneak-paths, and process variations.
RxNN [19]	C++	✓	✓	✓	Non-volatile memories.	Interconnects and sense parasitics, driver resistances, sneak paths, synaptic conductance variation.
NIXSim [37]	Python	✓	✗	✗	Non-volatile memories.	IR drop, SAF, programming/local/global variability, read noise.
PytorX [18]	Python	✓	✗	✓	Non-volatile memories.	IR drop, SAF, thermal noise, shot and random telegraph noise.

underlying memristor but also those of the implemented transistors within the crossbar array and the requisite peripherals.

Conversely, SySCIM [42] introduces a system-level simulator implemented in C++ and leveraging the SystemC and SystemC-AMS library. This approach empowers SySCIM to simulate an entire system encompassing a processor and the memristor-based computing core, setting it apart from the aforementioned frameworks. The usage of the SystemC-AMS timed data flow model enables continuous-time simulation of the crossbar, linking it to the event-driven simulation of the overall system. Furthermore, Device-to-Device (D2D) and Cycle-to-Cycle (C2C) variations are applicable through dedicated random or systematic variation functions for the fitting parameters of the memristive device. The memristor model offers 11 fitting parameters to customize the behavior within the computing core. Alongside this detailed model, SySCIM also offers a behavioral simulation model, providing an abstract representation of the memristor. To simulate the impact of SAFs, a dedicated fault-map can be incorporated to define the probability of each memristor being faulty.

3.2   Hardware Security in the Era of Neuromorphic Computing

Machine learning applications are pivotal in addressing complex challenges, including autonomous driving and image recognition. As a result, machine learning accelerators become targets for malicious entities that seek to manipulate inference results or steal intellectual property by acquiring trained model parameters. Thus, the domain of neuromorphic computing increasingly emphasizes the importance of hardware security. This section presents an overview of existing literature on three prevalent forms of attacks: Hardware Trojans, side-channel attacks, and fault injection attacks.

3.2.1   Hardware Trojans

Nagarajan et al. [35] introduce a concept known as the Emerging NVM-based Trojan Trigger (ENTT), featuring two forms of Trojan triggers built on ReRAM technology: a delay-based ENTT and a voltage-based ENTT. These triggers are activated by repeatedly accessing a particular memory location (Ntr times).

For the delay-based variant, resistance changes resulting from persistent access to a certain address lead to increased path delays. The trigger mechanism is built around an AND gate with two inputs: *Branch T* and *Branch B*. *Branch T* is responsible for outputting an inverted signal with a prolonged ON phase, while *Branch B* delivers a standard signal that extends its ON phase after each pulse. Upon the Ntr threshold being met, *Branch B*'s signal is sufficiently delayed producing a glitch in the AND gate together with the signal of *Branch T*.

On the other hand, the voltage-based trigger utilizes a comparator to determine the gradual resistance change from targeted memory access. A trigger signal is emitted when the comparator's sensed resistance shift crosses a predefined voltage reference. The authors also note that due to the non-volatility and consequent resistance retention in ReRAM cells, sporadic access to different cells can help circumvent detection mechanisms designed to flag repeated access patterns.

Khan et al. [27] propose an NVM hardware Trojan which can be intentionally activated or deactivated. The design consists of two primary components: a trigger and a payload. The trigger exploits the high write currents typical of NVM cells by writing to a chosen address with a specific pattern, causing a ground bounce. This effect is used to incrementally charge a capacitor, activating the Trojan when a certain voltage threshold is surpassed. Khan et al. describe three distinct payloads:

- **Information Leakage:** This payload copies data from a designated memory cell to another cell controlled by the attacker. It operates by connecting the two cells via the same bit and source lines, with a transistor injected to the victim cell's word line. Activation of the transistor by the Trojan allows for the copying of data when the victim cell is accessed.
- **Read Failure:** To disturb the read process of an NVM cell, the process variation-dependent V_{clamp} voltage can be modified. Since the read circuit of an NVM cell must account for process variations, a V_{clamp} voltage generator is used to adjust the read voltage according to variations detected after manufacturing. These generators typically consist of a resistor ladder with equal resistors. Connected to a multiplexer, the circuit can calibrate the read voltage post-manufacturing. An injected circuit can intentionally alter the generator's output, causing disruption in the read process.
- **Read/Write Failure:** By introducing an additional N-Type Metal–Oxide Semiconductor (NMOS) switch, this payload can short either the bit line or the source line to ground, or connect them to a $V_{disturb}$ to introduce noise during read/write operations.

Both studies cast light on the inadequacy of traditional detection methods, such as failure analysis tools, automatic test pattern generation, and side-channel analysis, against these NVM-specific Trojans. The negligible power consumption when inactive and their capacity to merge with standard memory operations when active make these Trojans particularly challenging to detect. To combat this threat, the authors propose a set of countermeasures including address scrambling, the use of ECC, machine learning analysis of memory images, and modulation of temperature and voltage to help identify and neutralize potential Trojans.

3.2.2   Side-Channel Attacks

STT-RAMs are increasingly recognized as a viable alternative to SRAM-based caches, primarily due to their high density and low power consumption. Nevertheless, STT-RAMs are plagued by high and asymmetric read/write currents, which potentially enable malicious attacks aimed at information leakage. Khan et al. [26] delve into this issue by investigating a differential power analysis-based side-channel attack, specifically targeting the recovery of the secret key during an AES-128 execution. The system under attack comprises a microcontroller executing the Advanced Encryption Standard (AES) algorithm, interfaced with an STT-RAM-based Last Level Cache (LLC). The authors assume that due to the limited number of general-purpose registers, intermediate data from the cryptographic algorithm is likely stored in the LLC. Their attack methodology successfully retrieves 8 bytes of the key using a minimum of 800 traces, a notably higher number than required for SRAM, attributable to STT-RAM's lower signal-to-noise ratio.

Likewise, Wang et al. [44] explore the exploitation of power side-channels to extract the complete network architecture of DNN models. They meticulously analyze power traces across various layer types, sequences, output channels/feature sizes of convolutional and fully connected layers, and kernel sizes of convolutional layers. The introduction of a mixed-signal power simulator, with configurable hardware-level properties and a PyTorch interface for mapping pre-trained Neural Network (NN) models, is a highlight of their approach. By recording power data from both digital and analog peripherals, they create lookup tables for use during simulation. This attack model assumes the adversary's familiarity with the hardware implementation of the accelerator and control over the input and output pins of the chip, albeit without access to individual memory cells. Their investigation demonstrates the feasibility of systematically extracting all layers and reconstructing the full NN model using the proposed attack methodology.

Turning to attacks on digital LIM architectures, SCARE [12] is focused on reverse engineering LIM gates using power and timing side-channels. The authors base their assumption on the execution of LIM operations over two cycles—first executing the AND gates, followed by the OR gate. Prior to extracting information, it is necessary to acquire template current profiles either through foundry-calibrated simulations or by fabricating test chips. SCARE presents two distinct attack models: the first is most effective when the inputs to the Boolean function are direct, while the second, more generic model requires extensive reverse engineering efforts but do not necessitate access to the direct inputs. This research demonstrates the ability of SCARE to reveal critical details about the implemented Boolean functions and highlights its potential in compromising real-world implementations.

3.2.3 Fault Injection Attacks

The Rowhammer attack on DRAMs has revealed a significant vulnerability by enabling the intentional injection of faults into computing and memory systems. Khan et al. [24] further explore this issue, examining the susceptibility of STT-RAMs to row hammering attacks. Their research focuses on how ground bounce, induced by high write currents, can compromise the integrity of STT-RAMs. This effect reduces the thermal energy barrier of memory bit cells, making them more prone to retention failures, magnetic field interferences, and thermal noise. A key finding of their study is that persistent writing (hammering) to a specific memory location can substantially weaken the thermal barrier of nearby unselected bits, resulting in bit flips. Moreover, the study indicates that ground bounce can impact bit-line and source-line drivers, negatively affecting the performance of selected cells by reducing the headroom voltage. Simulations in their research demonstrate that row hammering attacks can modify bits in STT-RAM within approximately 30.84 s, with the risk increasing at higher temperatures.

The same authors also investigate fault injection attacks on ReRAM-based caches, specifically for initiating DoS attacks [25]. They consider a 1T1R ReRAM-based LLC used by both the victim and the attacker. An attacker can write specific data patterns to produce deterministic supply noise in their memory space, facilitating DoS attacks or targeted polarity fault injection attacks on the shared memory space. The simulations show that attackers can launch DoS attacks by injecting over 120 mV of supply noise at the victim's write location, and a polarity fault injection attack with noise levels above 50 mV but below 120 mV. To counter these attacks, the authors suggest design-level countermeasures such as sequential read/write access, high-quality power/ground grids, and separate power rails for each bank.

Li et al. [30] present two innovative hardware attacks, Variation-oriented Adversarial Attack (VADER) and Enhanced Fault Injection Attack (EFI), exploiting the variability of ReRAMs. They aim to target a neuromorphic DNN accelerator based on memristive crossbar arrays. VADER employs a variation amplification algorithm to manipulate variation-sensitive pixels in input images, bypassing conventional defense mechanisms. This algorithm selects pixels in input samples that can magnify the effects of ReRAM crossbar variations, while limiting the number of affected pixels to conceal the manipulations. Conversely, EFI leverages ReRAM variations for covert and efficient low-cost fault injection attacks. It exploits variation-induced deviations in weight parameters to deliberately cause misclassification in specific sample categories by the NN model, minimizing the number of targeted weights to evade detection. Both attacks have demonstrated high success rates, underscoring the vulnerability of RRAM-based computing systems to hardware-aware adversarial attacks.

3.3   Instrumentation Platforms

Due to the immature state of memristive devices, various testing and instrumentation platforms have been proposed to evaluate single devices and crossbar structures.

Berdan et al. [4] developed the Memristor Characterization And Testing (mCAT) platform, designed to redistribute sneak-path currents within the crossbar array, significantly improving measurement accuracy. This platform provides all required potentials to the passive memristive crossbar through a multiplexer array. The bias generator applies the appropriate $V/2$ potentials to all unselected lines, while the sense bank aims to minimize measurement errors. Notably, the platform employs five different sense resistances, conducting 50 consecutive read operations with each resistance to diminish noise. Tested on a 32×32 discrete resistive crossbar array and solid-state TiO2-x ReRAM arrays, the mCAT platform demonstrated measurement accuracy with less than 1% error for standalone memristive devices and less than 10% error for 90% of devices in a custom resistive crossbar. Its versatility, enhanced by an NXP mBED microcontroller and a MATLAB Graphical User Interface (GUI), allows for seamless integration with ReRAM crossbar arrays. The mCAT platform, distributed commercially by ArC Instruments Ltd. under the name ArC ONE measurement system [3], features open-source software for the microcontroller and Python software for the host PC, including predefined test routines and data visualizations.

The mCAT system, however, is constrained by its interface to facilitate only passive (selectorless) memristive crossbar arrays. Addressing this, Foster et al. [14, 15] proposed an improved version centered around an FPGA EFM-03 development board. This system includes a 64-channel Source-Meter Unit (SMU) and two 32-pin banks for digital I/O, achieving a current noise floor of 170 pA, pulse delivery of ± 13.5 V, and a maximum current drive of 12 mA per channel. The SMU channel, a key subsystem, incorporates a programmable gain Transimpedance Amplifier (TIA) , a high-speed independent pulse generator, and a switch for current source access. The TIA functions as both a source and a meter, with differential ADCs for voltage reading. Digital terminals include a selector bank with 32 digital outputs and an arbitrary level logic bank supporting a wide voltage range. This system, capable of controlling up to a 32×32 selectorless crossbar or a 21×21 array with transistor selectors, is available from ArC Instruments Ltd. as the ArC TWO measurement system [3].

Kaya et al. [23] introduced another FPGA-based measurement system, focusing on 1T1R ReRAM structures. This system stands out for its straightforward and independent design, avoiding complex carrier modules. It addresses key ReRAM parameters like switching voltages, resistance states, data retention, and endurance. Based on a XILINX Artix 7 series FPGA board (AC701), it features a 5-channel arbitrary waveform generator and voltage buffers. The system supports both current and voltage-based resistance measurements, with a maximum output of 10 mA. The proposed platform utilizes a 32-Bit Microblaze soft processor with FreeRTOS, enabling flexible application development. The implemented software facilitates various memory operations and is complemented by a MATLAB GUI for cell operations and resistance variability analysis, useful for developing security applications like Random Number Generations (RNGs) and Physical Unclonable Functions (PUFs).

aixACCT Systems GmbH developed the aixMATRIX, a matrix test system designed for parallel stimulation of test structures on 64 analog channels. Each channel is equipped with a 16-bit DAC and supports a sample rate of 100 MS/ms. The system generates bipolar signals up to ± 10 V and includes ultrafast, bipolar adjustable current limiting. It is optimized for the analysis of memristive memories with response times below 50 ns. Both active and passive memory arrays up to 32×32 cells can be examined. The system is integrated into a 200 mm wafer prober to support research on neuromorphic memory systems [1].

3.4   Lessons Learned

The postulation of the memristor launched an avalanche of research to investigate these devices from the material stack to the system level performance and energy efficiency. The merit of works has focused on the analog CIM paradigm which promises to drastically outperform conventional von Neumann architectures. However, specifically this operational mode significantly suffers from the high variability of the memristive devices limiting the scope of real-world applications. Consequently, a thorough investigation of the impact of reliability issues on LIM operations is missing. While neuromorphic systems are gaining momentum, an increasing number of publications prove the existence of hardware security vulnerabilities stemming from the unique characteristics of the underlying memristive device. This novel attack surface represents a significant threat to future neuromorphic computing systems and to the application executed on them. To investigate concerns like the reliability and hardware security threats, instrumentation platforms have been proposed offering different measurement accuracies, in/output channels, and characterization routines. Nevertheless, the vast majority of publications bases their investigation on simulated devices failing to reproduce the impact of variability and reliability concerns of memristive devices. To the best of our knowledge, there is currently no instrumentation platform capable of performing analog nor digital operations on memristive crossbar arrays with the aim to determine the impact of these concerns on their computational result.

Therefore, this book aims to provide a comprehensive analysis of reliability and hardware security concerns of neuromorphic systems. Furthermore, it aims to add crucial findings to usher the way towards trustworthy neuromorphic computing. Alongside other contributions, the upcoming chapters introduce the following:

1. **Fault Injection Platform:** A fault injection platform for LIM operations allowing a comprehensive examination of reliability, ranging from the individual memristor level

up to complete, real-world workloads. This platform enables a comparative analysis of various logic families regarding their resilience to faults and supports detailed investigations at the application level to identify the parameters that most critically affect potential workloads.

2. **NeuroHammer**: This represents a novel hardware security attack, enabling adversaries to deliberately manipulate bits in ReRAMs. NeuroHammer poses significant challenges to the integrity of neuromorphic systems.

3. **NeuroBreakoutBoard**: An innovative instrumentation platform designed to facilitate CIM operations on actual memristive crossbar arrays. It provides valuable insights into how reliability issues can affect CIM operations.

3.5   Synopsis

This chapter presents a detailed survey of existing research in neuromorphic computing, focusing on the reliability and hardware security of memristor-based systems. It explores the challenges in ensuring the reliability of ReRAMs, including strategies for fault detection and error correction. The chapter also discusses various simulation platforms that model system performance and reliability, accounting for device variability and faults. A significant portion is dedicated to hardware security, examining vulnerabilities to hardware Trojans, side-channel, and fault injection attacks in neuromorphic systems. Additionally, it reviews different instrumentation platforms for testing and evaluating memristive devices and crossbar structures.

References

1. aixACCT Systems GmbH: aixMATRIX – Lightning-Quick Testing Of Neuromorphic Memory Systems (2023). https://www.aixacct.com/testsysteme/paralleltestsysteme/aixmatrix. Accessed Nov 14, 2023

2. Al-Shaarawy, A., Amirsoleimani, A., Genov, R.: PRUNIX: non-ideality aware convolutional neural network pruning for memristive accelerators. In: IEEE International Symposium on Circuits AND Systems (ISCAS). IEEE (2022). https://doi.org/10.1109/iscas48785.2022.9937541

3. ARC Instruments Ltd.: Arc Instruments - High Performance Array Control Instruments (2023). https://www.whitehouse.gov/briefing-room/statements-releases/2022/08/09/fact-sheet-chips-AND-science-act-will-lower-costs-create-jobs-strengthen-supply-chains-AND-counter-china/. Accessed Nov 13, 2023

4. Berdan, R., Serb, A., Khiat, A., Regoutz, A., Papavassiliou, C., Prodromakis, T.: A u-controller-based system for interfacing selectorless RRAM crossbar arrays. IEEE Trans. Electron Devices **62**(7), 2190–2196 (2015). https://doi.org/10.1109/ted.2015.2433676

5. Chakraborty, I., Ali, M.F., Kim, D.E., Ankit, A., Roy, K.: GENIEx: a generalized approach to emulating non-ideality in memristive XBars using neural networks. In: ACM/IEEE Design Automation Conference (DAC). IEEE (2020). https://doi.org/10.1109/dac18072.2020.9218688

6. Chaudhuri, A., Chakrabarty, K.: Analysis of process variations, defects, and design-induced coupling in memristors. In: IEEE International Test Conference (ITC). IEEE (2018). https://doi.org/10.1109/test.2018.8624819

7. Chen, Y.X., Li, J.F.: Fault modeling and testing of 1T1R memristor memories. In: IEEE VLSI Test Symposium (VTS). IEEE (2015). https://doi.org/10.1109/vts.2015.7116247

8. Chen, P.Y., Peng, X., Yu, S.: Neurosim+: an integrated device-to-algorithm framework for benchmarking synaptic devices and array architectures. In: IEEE International Electron Devices Meeting (IEDM). IEEE (2017). https://doi.org/10.1109/iedm.2017.8268337

9. Chen, P.Y., Peng, X., Yu, S.: NeuroSim: a circuit-level macro model for benchmarking neuro-inspired architectures in online learning. IEEE Trans. Comput.-Aided Design Integr. Circuits Syst. 37(12), 3067–3080 (2018). https://doi.org/10.1109/tcad.2018.2789723

10. Chua, L.: Memristor-the missing circuit element. IEEE Trans. Circuit Theory 18(5), 507–519 (1971). https://doi.org/10.1109/tct.1971.1083337

11. Emara, A.S., Madian, A.H., Amer, H.H., Amer, S.H., Abdelhalim, M.B.: Testing of memristor ratioed logic (MRL) XOR gate. In: International Conference on Microelectronics (ICM). IEEE (2016). https://doi.org/10.1109/icm.2016.7847939

12. Ensan, S.S., Nagarajan, K., Khan, M.N.I., Ghosh, S.: SCARE: side channel attack on in-memory computing for reverse engineering. IEEE Trans. Very Large Scale Integr. Syst. 29(12), 2040–2051 (2021). https://doi.org/10.1109/tvlsi.2021.3110744

13. Feinberg, B., Wang, S., Ipek, E.: Making memristive neural network accelerators reliable. In: IEEE International Symposium on High Performance Computer Architecture (HPCA). IEEE (2018). https://doi.org/10.1109/hpca.2018.00015

14. Foster, P., Huang, J., Serb, A., Prodromakis, T., Papavassiliou, C.: An FPGA based system for interfacing with crossbar arrays. In: International Symposium on Circuits and Systems (ISCAS). IEEE (2020). https://doi.org/10.1109/iscas45731.2020.9180671

15. Foster, P., Huang, J., Serb, A., Stathopoulos, S., Papavassiliou, C., Prodromakis, T.: An FPGA-based system for generalised electron devices testing. Sci. Rep. 12(1) (2022). https://doi.org/10.1038/s41598-022-18100-3

16. Hamdioui, S., Taouil, M., Haron, N.Z.: Testing open defects in memristor-based memories. IEEE Trans. Comput. 64(1), 247–259 (2015). https://doi.org/10.1109/tc.2013.206

17. Hamming, R.W.: Error detecting and error correcting codes. Bell Syst. Tech. J. 29(2), 147–160 (1950). https://doi.org/10.1002/j.1538-7305.1950.tb00463.x

18. He, Z., Lin, J., Ewetz, R., Yuan, J.S., Fan, D.: Noise injection adaption: end-to-end ReRAM crossbar non-ideal effect adaption for neural network mapping. In: Design Automation Conference. ACM (2019). https://doi.org/10.1145/3316781.3317870

19. Jain, S., Sengupta, A., Roy, K., Raghunathan, A.: RxNN: a framework for evaluating deep neural networks on resistive crossbars. IEEE Trans. Comput-Aided Design Integr. Circuits Syst. 40(2), 326–338 (2021). https://doi.org/10.1109/tcad.2020.3000185

20. Kannan, S., Karri, R., Sinanoglu, O.: Sneak path testing and fault modeling for multilevel memristor-based memories. In: IEEE International Conference on Computer Design (ICCD). IEEE (2013). https://doi.org/10.1109/iccd.2013.6657045

21. Kannan, S., Karimi, N., Karri, R., Sinanoglu, O.: Detection, diagnosis, and repair of faults in memristor-based memories. In: IEEE VLSI Test Symposium (VTS). IEEE (2014). https://doi.org/10.1109/vts.2014.6818762

22. Kannan, S., Karimi, N., Karri, R., Sinanoglu, O.: Modeling, detection, and diagnosis of faults in multilevel memristor memories. IEEE Trans. Comput.-Aided Design Integr. Circuits Syst. 34(5), 822–834 (2015). https://doi.org/10.1109/tcad.2015.2394434

23. Kaya, Z.E., Tekin, S.B., Kalem, S.: Design of an FPGA-Based RRAM parameter measurement platform. In: International Conference on Industrial Technology (ICIT). IEEE (2018). https://doi.org/10.1109/icit.2018.8352386

24. Khan, M.N.I., Ghosh, S.: Analysis of row hammer attack on STTRAM. In: IEEE International Conference on Computer Design (ICCD). IEEE (2018). https://doi.org/10.1109/iccd.2018.00021
25. Khan, M.N.I., Ghosh, S.: Fault injection attacks on emerging non-volatile memory and countermeasures. In: International Workshop on Hardware and Architectural Support for Security and Privacy. ACM (2018). https://doi.org/10.1145/3214292.3214302
26. Khan, M.N.I., Bhasin, S., Yuan, A., Chattopadhyay, A., Ghosh, S.: Side-channel attack on STTRAM based cache for cryptographic application. In: IEEE International Conference on Computer Design (ICCD). IEEE (2017). https://doi.org/10.1109/iccd.2017.14
27. Khan, M.N.I., Nagarajan, K., Ghosh, S.: Hardware trojans in emerging non-volatile memories. In: Design, Automation & Test in Europe Conference & Exhibition (DATE). IEEE (2019). https://doi.org/10.23919/date.2019.8714843
28. Lammie, C., Azghadi, M.R.: MemTorch: a simulation framework for deep memristive crossbar architectures. In: IEEE International Symposium on Circuits AND Systems (ISCAS). IEEE (2020). https://doi.org/10.1109/iscas45731.2020.9180810
29. Lammie, C., Xiang, W., Azghadi, M.R.: Modeling and simulating in-memory memristive deep learning systems: an overview of current efforts. Array **13**, 100116 (2022). https://doi.org/10.1016/j.array.2021.100116
30. Li, B., Lv, H., Wang, Y., Chen, Y.: Security threat to the robustness of RRAM-based neuromorphic computing system. In: International Symposium on Smart Electronic Systems (iSES). IEEE (2022). https://doi.org/10.1109/ises54909.2022.00061
31. Lin, M.Y., Cheng, H.Y., Lin, W.T., Yang, T.H., Tseng, I.C., Yang, C.L., Hu, H.W., Chang, H.S., Li, H.P., Chang, M.F.: DL-RSIM: a simulation framework to enable reliable ReRAM-based accelerators for deep learning. In: ACM International Conference on Computer-Aided Design (2018). https://doi.org/10.1145/3240765.3240800
32. Liu, P., You, Z., Wu, J., Liu, B., Han, Y., Chakrabarty, K.: Fault modeling and efficient testing of memristor-based memory. IEEE Trans. Circuits Syst. I Regular Papers **68**(11), 4444–4455 (2021). https://doi.org/10.1109/tcsi.2021.3098639
33. Mao, M., Cao, Y., Yu, S., Chakrabarti, C.: Optimizing latency, energy, and reliability of 1T1R ReRAM through appropriate voltage settings. In: IEEE International Conference on Computer Design (ICCD). IEEE (2015). https://doi.org/10.1109/iccd.2015.7357125
34. Mao, M., Cao, Y., Yu, S., Chakrabarti, C.: Optimizing latency, energy, and reliability of 1T1R ReRAM through cross-layer techniques. IEEE J. Emerg. Sel. Top. Circuits Syst. **6**(3), 352–363 (2016). https://doi.org/10.1109/jetcas.2016.2547745
35. Nagarajan, K., Khan, M.N.I., Ghosh, S.: ENTT: a family of emerging NVM-based trojan triggers. In: IEEE International Symposium on Hardware Oriented Security and Trust (HOST). IEEE (2019). https://doi.org/10.1109/hst.2019.8740836
36. Peng, X., Huang, S., Luo, Y., Sun, X., Yu, S.: DNN+NeuroSim: an end-to-end benchmarking framework for compute-in-memory accelerators with versatile device technologies. In: IEEE International Electron Devices Meeting (IEDM). IEEE (2019). https://doi.org/10.1109/iedm19573.2019.8993491
37. Quan, C., Fouda, M.E., Lee, S., Lee, J.: Multi-fidelity nonideality simulation and evaluation framework for resistive neuromorphic computing. In: Asilomar Conference on Signals, Systems, AND Computers. IEEE (2022). https://doi.org/10.1109/ieeeconf56349.2022.10052098
38. Rasch, M.J., Moreda, D., Gokmen, T., Gallo, M.L., Carta, F., Goldberg, C., Maghraoui, K.E., Sebastian, A., Narayanan, V.: A flexible and fast PyTorch toolkit for simulating training and inference on analog crossbar arrays. In: IEEE International Conference on Artificial Intelligence Circuits and Systems (AICAS). IEEE (2021). https://doi.org/10.1109/aicas51828.2021.9458494
39. Ravi, V., Prabaharan, S.R.S.: Memristor based memories: defects, testing, and testability techniques. Far East J. Electron. Commun. **17**(1), 105–125 (2017). https://doi.org/10.17654/ec017010105

40. Roy, S., Sridharan, S., Jain, S., Raghunathan, A.: TxSim: modeling training of deep neural networks on resistive crossbar systems. IEEE Trans. Very Large Scale Integr. Syst. **29**(4), 730–738 (2021). https://doi.org/10.1109/tvlsi.2021.3063543

41. Schechter, S., Loh, G.H., Strauss, K., Burger, D.: Use ECP, not ECC, for hard failures in resistive memories. In: ACM International Symposium On Computer Architecture (2010). https://doi.org/10.1145/1815961.1815980

42. Shadmehri, S.H.H., BanaGozar, A., Kamal, M., Stuijk, S., Afzali-Kusha, A., Pedram, M., Corporaal, H.: SySCIM: SystemC-AMS simulation of memristive computation in-memory. In: Design, Automation & Test in Europe Conference & Exhibition (DATE). IEEE (2022). https://doi.org/10.23919/date54114.2022.9774749

43. Staudigl, F., Merchant, F., Leupers, R.: A survey of neuromorphic computing-in-memory: architectures, simulators, and security. IEEE Design Test **39**(2), 90–99 (2022). https://doi.org/10.1109/mdat.2021.3102013

44. Wang, Z., hsuan Meng, F., Park, Y., Eshraghian, J.K., Lu, W.D.: Side-channel attack analysis on in-memory computing architectures. IEEE Trans. Emerg. Top. Comput. 1–13 (2023). https://doi.org/10.1109/tetc.2023.3257684

45. Xia, L., Li, B., Tang, T., Gu, P., Chen, P.Y., Yu, S., Cao, Y., Wang, Y., Xie, Y., Yang, H.: MNSIM: simulation platform for memristor-based neuromorphic computing system. IEEE Trans. Comput.-Aided Design Integr. Circuits Syst. 1–1 (2017). https://doi.org/10.1109/tcad.2017.2729466

46. Xia, L., Huangfu, W., Tang, T., Yin, X., Chakrabarty, K., Xie, Y., Wang, Y., Yang, H.: Stuck-at fault tolerance in RRAM computing systems. IEEE J. Emerg. Sel. Top. Circuits Syst. **8**(1), 102–115 (2018). https://doi.org/10.1109/jetcas.2017.2776980

47. Xiao, T.P., Bennett, C.H., Feinberg, B., Marinella, M.J., Agarwal, S.: CrossSim: Accuracy Simulation of Analog In-Memory Computing. https://github.com/sANDialabs/cross-sim

48. Xu, C., Niu, D., Zheng, Y., Yu, S., Xie, Y.: Impact of cell failure on reliable cross-point resistive memory design. ACM Trans. Design Automation Electron. Syst. **20**(4), 1–21 (2015). https://doi.org/10.1145/2753759

49. Zhang, Y., Feng, D., Tong, W., Hua, Y., Liu, J., Tan, Z., Wang, C., Wu, B., Li, Z., Xu, G.: CACF: A novel circuit architecture co-optimization framework for improving performance, reliability and energy of ReRAM-based main memory system. ACM Trans. Archit. Code Optim. **15**(2), 1–26 (2018). https://doi.org/10.1145/3195799

50. Zhang, B., Uysal, N., Fan, D., Ewetz, R.: HANDling stuck-at-faults in memristor crossbar arrays using matrix transformations. In: ACM Asia and South Pacific Design Automation Conference (2019). https://doi.org/10.1145/3287624.3287707

51. Zhang, Y., Yu, Z., Gu, L., Wang, C., Feng, D.: EnTiered-ReRAM: an enhanced low latency and energy efficient TLC crossbar ReRAM architecture. IEEE Access **9**, 167173–167189 (2021). https://doi.org/10.1109/access.2021.3129878

52. Zheng, Y., Xu, C., Xie, Y.: Modeling framework for cross-point resistive memory design emphasizing reliability and variability issues. In: IEEE Asia and South Pacific Design Automation Conference (2015). https://doi.org/10.1109/aspdac.2015.7058990

53. Zhu, Z., Sun, H., Qiu, K., Xia, L., Krishnan, G., Dai, G., Niu, D., Chen, X., Hu, X.S., Cao, Y., Xie, Y., Wang, Y., Yang, H.: MNSIM 2.0: a behavior-level modeling tool for memristor-based neuromorphic computing systems. In: Great Lakes Symposium on VLSI. ACM (2020). https://doi.org/10.1145/3386263.3407647

Fault Injection in Logic-in-Memory Architectures

4

As discussed in Chap. 2, there are two fundamental ways to perform computation with memristive crossbar arrays. Analog Computing-in-Memory (CIM) leverages the continuous resistance values of memristive devices to perform Multiply–Accumulate (MAC) operations in the analog domain. While offering exceptional performance through extensive parallel computation, analog CIM copes with inherent limitations, such as the need for Digital-to-Analog Converters (DACs)/Analog-to-Digital Converters (ADCs) for value conversion between digital and analog domains. On the contrary, Logic-in-Memory (LIM) is another flavor, executing logic functions within memristive crossbar arrays by strictly utilizing the memristive devices in a binary manner. This approach may reduce computing performance compared to analog CIM but promises enhanced reliability without requiring conversion between the analog and digital domains. However, as concluded in Chap. 3, the reliability of LIM architectures remains largely unexplored, with the majority of research focusing on analog CIM. LIM's reliance on logic families to implement LIM gates introduces an additional layer of complexity and uncertainty by potentially utilizing a faulty memristor multiple times within a single logic gate. Consequently, modeling the reliability of such operations requires a fundamentally different approach that accounts for the memristive devices, the logic family, and the executed application. Therefore, this chapter introduces a comprehensive fault injection framework for LIM operations, capable of investigating reliability from the memristor level to actual full-fledged workloads. The framework includes a memristor-level fault injection simulator named **X-Fault** and an operational-level simulator called **FLIM** . The former excels at emulating the impact of faults on individual logic gates with high precision, while the latter offers exceptional simulation speed for executing realistic workloads using abstracted fault models. Together, these simulators allow for a comparison of logic families in terms of fault resilience

© The Author(s), under exclusive license to Springer Nature Switzerland AG 2026
F. Staudigl, R. Leupers, *Towards Trustworthy Neuromorphic Computing*,
Synthesis Lectures on Engineering, Science, and Technology,
https://doi.org/10.1007/978-3-032-09586-2_4

and enable an investigation at the application level to understand which parameters most significantly influence potential workloads.

This chapter is organized as follows: Sect. 4.1 provides an overview of the framework. The process of application mapping is discussed in Sect. 4.2, while Sect. 4.3 outlines the crossbar simulator and its fault models. Delving into the internals of FLIM , Sect. 4.4 elaborates on the fault generation mechanism, and Sect. 4.5 discusses the implementation of the fault injection method. Comprehensive case studies presented in Sect. 4.7 underscore the significance of our fault injection framework. Finally, in Sect. 4.8, we discuss the limitations and future directions of this work.

4.1 Framework Overview

An overview of the fault injection framework is depicted in Fig. 4.1. This framework consists of two separate modules, X-Fault and FLIM , specifically designed for the execution of Binary Neural Networks (BNNs), which predominantly utilize binary XNOR operations. X-Fault requires inputs such as hardware constraints, fault type, and the type of logic gate. The hardware constraints specify the dimensions of the crossbar array and the mapping algorithm, while the fault type identifies the fault model and its parameters, namely, injection rate and fault pattern. The logic gate parameter not only determines the gate itself but also the corresponding logic family. In general, X-Fault comprises a mapping tool and a crossbar model. The mapping tool can interpret Larq [8] models and relay the relevant XNOR operations to the crossbar model while all other operations are executed within TensorFlow [1]. The crossbar simulator incorporates a comprehensive memory controller which orchestrates the instantiated crossbar arrays. Closely linked with

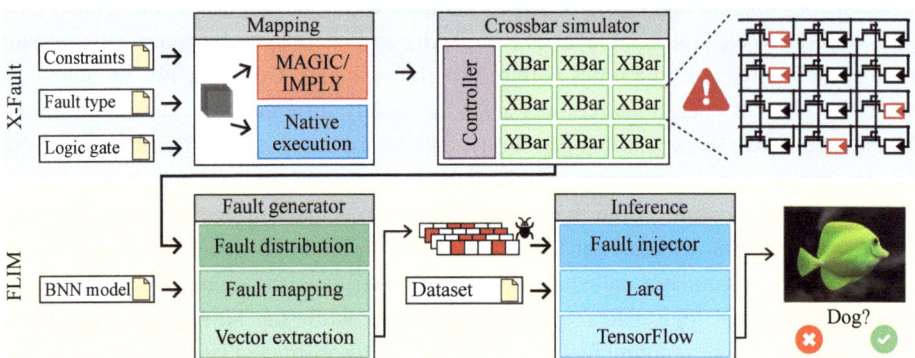

Fig. 4.1 Overview of the fault injection framework featuring X-Fault and FLIM: X-Fault provides a mapping tool and crossbar simulator for highly accurate simulations that assess the resilience of logic families. In contrast, FLIM employs a more abstracted approach with its fault generator and fault injector, offering a platform for high-speed simulation

the crossbar model is the binary memristor model, which implements various fault models that can be parameterized to accommodate a wide array of experiments [17–19].

The second component of the simulation framework, FLIM , employs an abstracted fault injection methodology aimed at achieving high simulation speed. This simulator requires a BNN model based on the Larq framework. The necessary fault distribution can be sourced from X-Fault or input manually. To significantly boost simulation speed, FLIM transfers the labor-intensive tasks to a preprocessing phase. The fault generator embodies these preprocessing steps, comprising three distinct phases: fault distribution, mapping, and vector extraction. The resultant noise vector serves as an input for the fault injector, which is deeply integrated within the TensorFlow/Larq environment.

4.2   Application Mapping

X-Fault's mapping tool adeptly maps convolutional and dense layers of a BNN onto a crossbar of a predefined size. The mapping tool extracts the dimensions, as well as the respective values of the kernel and inputs, to generate a set of write, read, and logic instructions. Central to the mapping algorithm is its focus on minimizing the number of instructions. This is achieved by efficiently tracking partial fits of the kernel within the remaining cells of a crossbar array. The mapper facilitates communication with the crossbar simulator through a binary file interface, leveraging the FlatBuffers library [9] which offers an efficient and cross-platform serialization methodology. The binary file encapsulates the memory addresses of the kernels, their corresponding values, and the requisite logic operations.

At its core, the algorithm adopts a weight stationary mapping, wherein the weights (kernels) of the BNN are sequentially mapped onto the crossbar array and remain stationary throughout the computation process. While the simulator does not account for weight drift effects, it is noteworthy that depending on the logic family used, some input cells may be overwritten during the execution of logic gates, necessitating a rewrite of weights. Algorithm 4.1 shows the implemented mapping methodology, which is detailed in the following.

Initially, the algorithm calculates the total number of values in the kernel matrix and processes these values in segments iteratively. In each iteration, a particular kernel segment, determined by the number of XNOR gates in the crossbar, is selected and stored in the binary file. These segments are computed based on the total number of kernels multiplied by the number of values per kernel.

Next, the algorithm identifies the starting and ending points of the kernel within the input matrix. This step is critical for accurately aligning the kernel with the input data for effective XNOR operations. To achieve this, nested loops are utilized to traverse the dimensions of the input matrix, tailored to the size of the kernel. Within these loops, the algorithm linearly maps the input values to a one-dimensional list object, corresponding to the kernel values. This structured input list is then written to the binary file, aligned with the

---

Algorithm 4.1: Weight stationary XNOR mapping on memristive crossbar arrays

---

Data: crossbar instance (*crossbar*), kernel instance (*kernel*), input data instance (*input*),
         FlatBuffer handler instance (*flatbuffer_handler*)
Result: Mapped XNOR operations on crossbar array

1 *nr_kernel_values* ← *kernel*.nr_values_per_kernel × *kernel*.nr_kernels;
2 **for** *i* ← 0 **to** *kernel.kernel_iterations()* **do**
3 | *kernel_to_write* ← *kernel*.get_kernel_index(i,*nr_xnor_gates*);
4 | *crossbar*.write_kernel(*kernel_to_write*, *flatbuffer_handler*);
5 | *kernel_start* ← i × *nr_xnor_gates*;
6 | *kernel_end* ← (i + 1) × *nr_xnor_gates*;
7 | **if** *kernel_end* > *nr_kernel_values* **then**
8 | | *kernel_end* ← *nr_kernel_values*;
9 | **end**
10 | (*i_values*, *i_indices*) ← get_input(*kernel_start*, *kernel_end*);
11 | **for** *y* ← 0 **to** *input.get_size()[0]* − *kernel.get_size()[0]* +1 **do**
12 | | **for** *x* ← 0 **to** *input.get_size()[1]* − *kernel.get_size()[1]* +1 **do**
13 | | | *crossbar_input* ← empty list;
14 | | | **for** *each pair (input, input_loc) in zip(i_values, i_indices)* **do**
15 | | | | *input_list* ← list containing int(*input*);
16 | | | | *y_p* ← *input_loc[1]* + y;
17 | | | | *x_p* ← *input_loc[0]* + x;
18 | | | | *crossbar_input_element* ← list containing *input_list, y_p, x_p*;
19 | | | | append *crossbar_input_element* to *crossbar_input*;
20 | | | **end**
21 | | | *crossbar*.write_input_to_crossbar(*crossbar_input*, *flatbuffer_handler*);
22 | | | *crossbar*.calculate_xnor(*flatbuffer_handler*);
23 | | **end**
24 | **end**
25 **end**

---

kernel parameters. Upon successful mapping of inputs and kernels, the mapping tool issues a compute command, activating the crossbar simulator to execute the XNOR operations as per the instructions in the binary file. This algorithmic design not only maximizes the parallel computational capacity of the crossbar array but also aligns with the constraints of the implemented LIM families.

4.3   Crossbar Simulator

The crossbar simulator follows the hierarchy of Dynamic Random-Access Memory (DRAM) consisting of channels, ranks, banks, rows, and columns. Internally, the simulator implements this hierarchy in three distinct entities: memory controller, crossbar, and memristor. In the following, the functionality of each entity is described in detail.

4.3.1 Memory Controller

The memory controller parses the provided binary file from the mapping tool to extract the required kernels and input values. Furthermore, the controller is responsible for orchestrating the underlying crossbars by issuing the instructions to the respective banks. The overall dimensions of the memory can be customized via a configuration file which determines the number of ranks, banks, and the dimensions of the crossbar arrays. The memory address translation and allocation employs two different interleaving methods (page or rank interleave) to manage how physical addresses are mapped to memory channels, ranks, banks, rows, and columns. This flexibility allows for a more comprehensive exploration of memory management strategies in the presence of faults. Additionally, the implementation takes into account the bit-level manipulation of data, foster the unique requirements of LIM.

4.3.2 Crossbar Model

The crossbar model is designed to simulate a binary memristive crossbar array, with a particular focus on the implications of in-field faults within the system. It abstracts the interactions between memristive cells, aiming to balance the accuracy of simulation with the need for high simulation speed. Each cell within the array is instantiated through the memristor model.

Initialization of the crossbar is executed by parsing a configuration file that determines the crossbar's dimensions, the proportion of memristors in the Low Resistive State (LRS) versus the High Resistive State (HRS), the prevalence of faulty cells, and the characteristics of faults, including the type and pattern. To assign initial states and fault models to cells, the crossbar model employs a uniformly distributed random number generator. The distribution of these random values is given by the probability density function:

$$P(x|a, b) = \frac{1}{b - a} \tag{4.1}$$

Here, x represents the random value uniformly distributed over the interval $[a, b)$. In addition to cell initialization, the crossbar constructs inter-cell relationships to account for coupling faults. Read and write operations are conducted on a per-row basis. While the model incorporates standard read and write methods, it also features a specialized interface to support access based on the $V/2$ scheme, as discussed in Sect. 2.1.2.

Table 4.1 presents a comprehensive overview of the logic gates implemented within the simulation framework. Both the Memristor-Based Material Implication (IMPLY) and Memristor-Aided Logic (MAGIC) logic families offer only a fundamental set of logic gates. To construct more complex Boolean functions, the simulator combines these basic gates, which, as a result, significantly increases the number of required cycles and

Table 4.1 Overview of logic gate implementations based on IMPLY and MAGIC logic families including the number of memristors (#mem) and cycles (#cycles) required for each operation

Logic family	Logic gate	#mem	#cycles	Internal structure
IMPLY	IMP	2	1	IMP(1,2)
	NOT	2	2	FILL(2,HRS) →IMP(1,2)
	NAND	3	3	FILL(3,HRS) →IMP(2,3) →IMP(1,3)
	AND	3	4	NAND(1,2,3) →NOT(3,1)
	OR	3	3	FILL(3,HRS) →IMP(1,3) →IMP(3,1)
	NOR	3	5	OR(1,2,3) →NOT(2,1)
	XOR	4	5	IMP(1,3) →IMP(4,2) →NOT(2,4) →IMP(3,4)
	XNOR	4	6	XOR(1,2,3,4) →NOT(4,1)
MAGIC	NIMP	3	1	NIMP(1,2,3)
	NOR	3	1	NOR(1,2,3)
	OR	3	1	OR(1,2,3)
	XOR	3	3	FILL(3,HRS) →NIMP(1,2,3) →NIMP(2,1,3)
	XNOR	4	6	FILL(4,LRS) →FILL(3,HRS) →NIMP(1,2,3) →NIMP(2,1,3) →FILL(1,HRS) →NIMP(4,3,1)
	AND	5	9	OR(1,2,3) →FILL(4,HRS) →NIMP(1,2,4) →FILL(5,HRS) →NIMP(3,4,5) →FILL(4,HRS) →NIMP(2,1,4) →FILL(1,HRS) →NIMP(5,4,1)
	NAND	5	12	AND(1,2,3,4,5) →FILL(2,HRS) →FILL(3,LRS) →NIMP(3,1,2)

memristors, as detailed in Table 4.1. To streamline the implementation process of these logic gates, we introduce a utility function as follows:

Definition 4.1 The FILL(row, value) function assigns a specified *value* (either HRS or LRS) to a given *row*.

The implemented memory controller mandates that read and write operations are executed row-wise. Therefore, logic gates are mapped vertically within the crossbar array. As discussed in Sect. 2.2.2, the execution of a single logic operation triggers computation across all available columns, resulting in substantial parallelization. The internal structure shown in Table 4.1 refers to the rows using unique integer numbers ranging from $[1, n]$ with $n \in \mathbb{N}$ where n denotes the total number of rows of the crossbar. For example, the NAND gate constructed using the IMPLY logic family requires three memristors in three adjacent rows. Initially, the third row is set to HRS. Subsequently, Material Implication (IMP) operations are executed between the second and third rows, and then between the first and third rows, resulting in the output of the NAND gate, which utilizes three computational cycles/steps.

4.3.3 Memristor Model

The lowest level of X-Fault's memory hierarchy is represented by the memristor model which encapsulates the essential characteristics and functionalities of a memristive device. The abstracted model tracks the utilization by accumulating the read, write, and $V/2$ accesses. The simulator implements the MAGIC and IMPLY logic family because of their sole use of the memristive crossbar without any required peripherals, except the required resistor for the IMPLY family which we omit in our investigations. Furthermore, the memristor model utilizes an action log to keep a history of executed instructions for each cell. This history is required to facilitate dynamic faults which occur every n-th operation to be sensitized [10]. The memristor model only implements the traditional fault models outlined in Sect. 2.3.3, as the unique fault models require more precise simulations due to their analog characteristics. The fault models are randomly assigned to a certain percentage of the instantiated memristors and are enumerated as follows:

1. **Stuck-at-Fault (SAF)** is characterized by the memristor adopting a constant resistive value, manifesting either as a HRS or an LRS.
2. **Read-Destructive-Fault (RDF)** inverts the current state of the memristive cell while still returning a correct value.
3. **Deceptive Read Destructive Fault (DRDF)** modifies the current state of the cell and yields an incorrect value, misleading the read operation.
4. **Incorrect Read Fault (IRF)** causes the cell to remain unchanged, but the value retrieved is incorrect.
5. **Slow Write Fault (SWF)** occurs when a write operation to the cell fails, resulting in the preservation of the cell's previous state.
6. **Coupling Fault (CF)** occur when a write operation to a cell results in an unintended write operation to an adjacent cell.

Algorithm 4.2 illustrates the function used to evaluate the state of a cell in the presence of potential coupling faults induced by surrounding aggressors. The algorithm iteratively examines each cell based on a defined coupling pattern, utilizing directional pointers (east, west, north, south) to navigate the crossbar array. If the pattern is absent or the target cell's neighbors do not match the aggressor state, the algorithm terminates prematurely, indicating that no coupling fault will occur. However, if the aggressor pattern is detected, the state of the target memristor is toggled between the LRS and HRS, simulating the read/write disturb fault.

Algorithm 4.2: Coupling fault write/read disturb algorithm

Data: pattern length (*pattern_len*), coupling pattern (*pattern*), memristor pointers (*east*, *west*, *north*, *south*)

Result: state of the memristor cell after evaluating the coupling fault condition

1 **for** $i \leftarrow 0$ **to** *pattern_len* $- 1$ **do**
2 $\quad$ *it_ptr* $\leftarrow$ *this*;
3 $\quad$ **for** $j \leftarrow 0$ **to** $|pattern[i].x|$ - *1* **do**
4 $\quad\quad$ **if** *pattern[i].x* ≥ 0 **then**
5 $\quad\quad\quad$ *it_ptr*$\rightarrow$east $\neq$ NULL? (*it_ptr* $\leftarrow$ *it_ptr*$\rightarrow$east) : exit;
6 $\quad\quad\quad$ *it_ptr*$\rightarrow$west $\neq$ NULL? (*it_ptr* $\leftarrow$ *it_ptr*$\rightarrow$west) : exit;
7 $\quad\quad$ **end**
8 $\quad$ **end**
9 $\quad$ **for** $j \leftarrow 0$ **to** $|pattern[i].y|$ - *1* **do**
10 $\quad\quad$ **if** *pattern[i].y* ≥ 0 **then**
11 $\quad\quad\quad$ *it_ptr*$\rightarrow$north $\neq$ NULL? (*it_ptr* $\leftarrow$ *it_ptr*$\rightarrow$north) : exit;
12 $\quad\quad\quad$ *it_ptr*$\rightarrow$south $\neq$ NULL? (*it_ptr* $\leftarrow$ *it_ptr*$\rightarrow$south) : exit;
13 $\quad\quad$ **end**
14 $\quad$ **end**
15 $\quad$ **if** *it_ptr*$\rightarrow$get_state() $\neq$ *pattern[i].aggressor_state* **then**
16 $\quad\quad$ exit;
17 $\quad$ **end**
18 **end**
19 *this*$\rightarrow$state $\leftarrow$ (*this*$\rightarrow$state = LRS ? HRS : LRS);

4.4   Fault Generator

As discussed in Sect. 4.1, the FLIM simulator is designed to significantly enhance the simulation speed by abstracting the fault injection methodology. Specifically, this simulator separates the computationally intensive aspects of the fault injection, performing them offline, while the core fault injection process is integrated with the inference phase. This preprocessing stage, referred to as the fault generator, encompasses three primary steps: fault distribution, fault mapping, and noise vector extraction.

4.4.1   Fault Distribution

In contrast to X-Fault, FLIM introduces faults at the level of XNOR operations to enhance simulation performance, as depicted in Table 4.2. This abstraction level prohibits the ability to simulate single-cell accesses, thereby limiting the emulation of complex fault models. FLIM , therefore, restricts the supported fault models to primarily stuck-at and bit-flip faults.

In FLIM , the emulation of a crossbar array is bypassed in favor of annotating the inference process with fault masks. Figure 4.2 illustrates the correlation between randomly

Table 4.2 Comparison of fault models in X-Fault and FLIM: the fault models handled by X-Fault with their corresponding abstracted representations in FLIM illustrates the differences in fault handling between the two simulators and emphasizes the trade-off between simulation accuracy and speed

X-Fault	FLIM
Stuck-at fault	Stuck-at mask
Read destructive fault	Bit-flip mask
Deceptive read destructive fault	
Incorrect read fault	
Slow write fault	
Coupling fault	
Dynamic faults	Repeated bit-flip mask

assigned values in fault masks and the potential fault distribution within a crossbar array. This is illustrated in Fig. 4.2a, where in-field faults lead to malfunctioning XNOR operations. The FLIM fault masks, as shown in Fig. 4.2b–c, denote the locations of these malfunctions and are mapped onto the workload during the fault mapping stage.

The bit-flip mask is a two-dimensional Boolean array initially filled with zeros. The array's fault distribution, dictated by the injection rate, is represented by assigning a corresponding number of elements to one. FLIM also allows for the analysis of faults impacting entire rows or columns, stemming from potential address decoder issues. In these cases, the affected row or column in the bit-flip mask is entirely marked as one. FLIM is capable of handling dynamic faults, which necessitate the duplication of the fault mask across multiple layers. This replication involves constructing a sequence of bit-flip masks, each sequentially applied to different layers during the inference stage. Similarly, the stuck-at mask is represented as a two-dimensional Boolean array, initialized with zeros and stuck-at faults marked with ones.

This pre-processing method of generating masks significantly enhances performance by shifting the computationally intensive tasks of mapping and distributing faults away from the actual process of inference.

4.4.2   Fault Mapping

In the next phase, the masks previously generated are allocated to designated layers within the BNN model. For this purpose, the framework requires information about the dimensions and the total number of crossbars assumed to be utilized in the hardware accelerator. Initially, the mapping tool computes the number of concurrent XNOR operations based on the number of crossbars. As shown in Table 4.1, the IMPLY and MAGIC XNOR operation requires a total of four memristors representing the dominant Boolean function in BNNs.

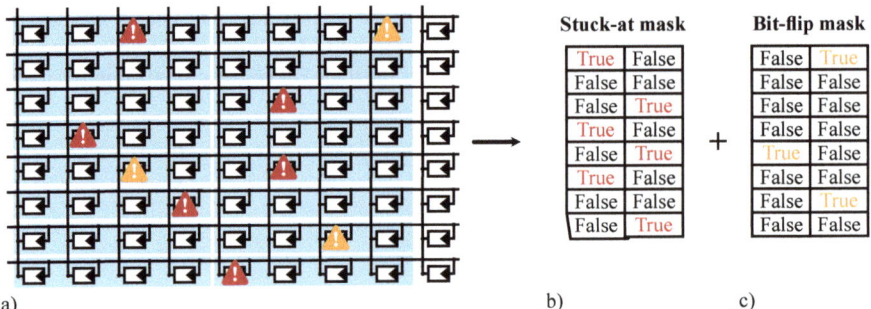

Fig. 4.2 Overview of the implemented fault mapping: detailed correlation between the presumed faulty memristive devices and the resulting stuck-at and bit-flip masks

Next, the tool extracts the number of required XNOR operations based on the provided model. Due to the structure of the BNN layers, XNOR operations are dominantly utilized in the 2-dimensional convolution layers and fully binarized dense layers. Therefore, these specific layers are targeted for mapping onto memristive crossbar arrays to leverage acceleration, while other layers continue to be processed using conventional Complementary Metal-Oxide Semiconductor (CMOS) technology. Hence, the mapping tool extracts the dimensions of these layers and assigns the previously generated fault masks.

4.4.3 Noise Vector Extraction

Finally, the generated fault masks are transformed in a 1-dimensional representation and stored in a binary file together with meta information about the target layer, the in-layer location, and the potential dynamic fault behavior. The binary file is independent of the dataset and allows for a parallel simulation of different system parameters to facilitate, for example, a reliability assessment of an architecture exploration.

4.5   Fault Injector

The Fault Injector stands as the core component of the FLIM platform, seamlessly integrated within the Larq and TensorFlow frameworks to achieve optimal performance through precise fault injection. Larq, essentially an extension of the Keras framework [4], is specifically designed to facilitate BNNs by introducing custom quantized layers that enhance the capabilities of standard Keras layers. In this context, we have augmented the base class of these layers by incorporating an instance of the Fault Injector. This integration allows the activation of the fault injection mechanism during the inference process, necessitating a modification of the original convolution method. Figure 4.3 visually depicts FLIM's fault injection approach. As elaborated in Sect. 4.4, FLIM is

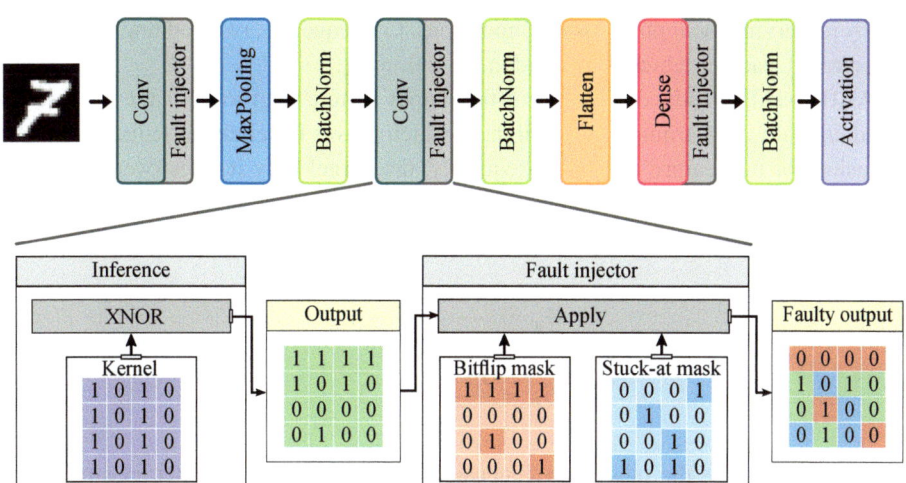

Fig. 4.3 Overview of FLIM's fault injection methodology composed of the inference stage and the fault injector

capable of injecting faults in both convolutional and dense layers. The process begins with the computation of the convolution between the feature map and the kernel, which initially remains fault-free, thereby yielding a correct computation outcome. Following this, the bit-flip and stuck-at masks are applied to the output through an additional XNOR operation, effectively integrating the predetermined faults based on the injection rate. The subsequent subsections provide a comprehensive exploration of the layer specific implementation of the fault injection mechanism.

4.5.1   Fault Injection in Conv2D Layers

The Larq library implements the `Quant2D` layer by taking advantage of the Tensor-Flow `tf.nn.conv2d()` function which performs a convolution of the input feature map and the given kernel. This function is invoked from the `convolution_op()` function within the Conv2D layer which acts as the entry point for FLIM's fault injector. The Fault Injector provides a custom convolution method extending the `tf.nn.conv2d()` function by injecting bit-flip and stuck-at faults.

Initially, the custom convolution function preprocesses the input feature map, tailoring its shape to align with the fault masks. This preprocessing is crucial because when a kernel is applied to the input, it typically results in the reduction of the spatial dimensions (width and height) of the output feature map. Such a reduction, particularly in deep networks, can lead to a significant decrease in input size and the potential loss of edge information. To mitigate this issue, padding techniques are employed to add extra pixels around the input image's border. This step ensures that the kernel can effectively process the border pixels, thereby preserving the input's spatial dimensions. Consequently, two padding methods are implemented, each of which is detailed in the following:

VALID padding reduces the size of the output feature map to omit the necessity to add additional values around the actual input which proves beneficial to reduce the spatial dimensions of the feature maps. Consequently, the kernel only iterates over a subset of pixels resulting in a reduced dimension of the output. The resultant output width O_w and height O_h is defined as

$$O_w = \frac{I_w - K_w}{s_w} + 1 \tag{4.2}$$

$$O_h = \frac{I_h - K_h}{s_h} + 1 \tag{4.3}$$

where I_w, I_h denotes the input data dimensions, K_w, K_h denotes the kernel dimensions, and s_w, s_h the width/height of the stride, respectively [7].

SAME padding adds additional pixels around the edges of the input image so that the output feature map matches the dimensions of the input data. The resultant output width O_w and height O_h is defined as

$$O_w = \frac{I_w + 2p - K_w}{s_w} + 1 \tag{4.4}$$

$$O_h = \frac{I_h + 2p - K_h}{s_h} + 1 \tag{4.5}$$

where I_w, I_h denotes the input data dimensions, K_w, K_h denotes the kernel dimensions, s_w, s_h the width/height of the stride, and p represents the number of zeros added along both axis [7].

Figure 4.4 illustrates the preprocessing required for the input image and the associated kernel in a convolutional layer. Patches corresponding to the kernel's size are extracted from the padded input and prepared for the XNOR operation with the kernel. Both the patches and the kernel are reshaped—flattened and duplicated—to enable the binary convolution. Following the reshaping, bit-flip and stuck-at masks are applied through additional XNOR operations, producing a feature map that incorporates the intended faults. This output is subsequently reshaped back to the form it would take after a standard, non-binary convolution. To finalize the output, TensorFlow's `reduce_sum` function aggregates values across a designated dimension, streamlining the tensor's structure. Each aggregated value thus forms a part of the concatenated output, with each contributing to the overall feature representation.

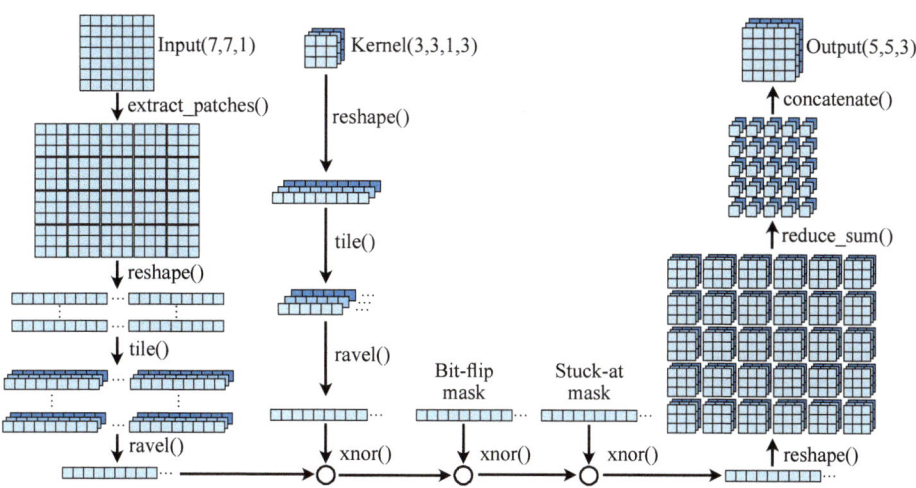

Fig. 4.4 Overview of the convolutional layer preprocessing including binary XNOR operation with fault injection and tensor aggregation

4.5.2 Fault Injection in Dense Layers

In comparison to convolution layers, dense layers, also known as fully connected layers, operate by performing a matrix multiplication between the input features and the weights of the neurons, followed by the addition of a bias term. Mathematically, the output $\mathbf{o}$ of a dense layer can be expressed as:

$$\mathbf{o} = f(\mathbf{W} \cdot \mathbf{x} + \mathbf{b}) \tag{4.6}$$

where $\mathbf{o}$ is the output vector, $\mathbf{W}$ is the weight matrix, $\mathbf{x}$ is the input vector, $\mathbf{b}$ is the bias vector, and f denotes the activation function. Since dense layers operate on a flattened input where the spatial structure is not preserved, padding is not required. Therefore, the input preprocessing of dense layers is simplified as shown in Fig. 4.5. First, the input, initially in a 2-dimensional form, is expanded by the `tile()` function, which replicates the tensor to match the kernel's dimensions. This is followed by `ravel()`, which flattens the expanded tensor into a 1-dimensional array. The kernel undergoes a similar `tile()` and `ravel()` process. Subsequently, both the input and kernel maps are subject to a XNOR operation resulting in the output feature map. Subsequently, the bit-flip and stuck-at masks are applied introducing faults in the layer. After the fault injection, the resultant feature map is reshaped, preparing it for the subsequent layer or output.

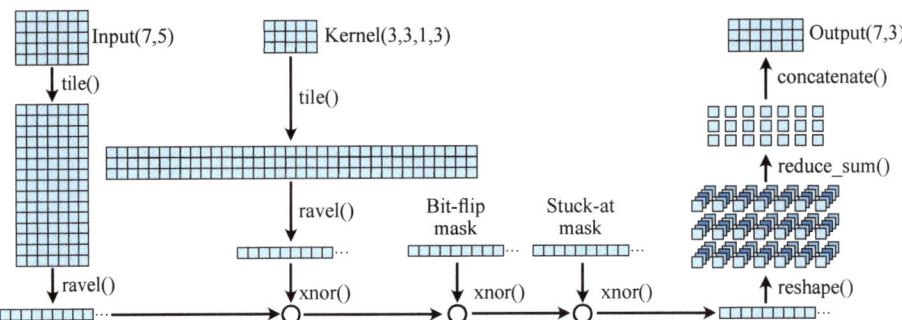

Fig. 4.5 Overview of the dense layer preprocessing including binary XNOR operation with fault injection and tensor aggregation

4.6   Resilience Metric for Logic-in-Memory Families

In the context of LIM architectures, a diverse set of logic families has been proposed, each with its unique characteristics and operational efficiencies. However, a critical aspect that often remains ambiguous is their resilience to faults, which is a key characteristic in terms of the real-world feasibility. This is particularly crucial as memristive devices, the backbone of these architectures, are inherently prone to various in-field faults. To address this gap, a standardized metric is essential for quantitatively assessing the resilience of these logic families and their individual gates. Our simulation framework, as detailed in Chap. 4, lays the groundwork for an in-depth analysis of the reliability of LIM gates. However, it lacks a universal metric that can effectively measure and compare the fault tolerance across different logic families. Such a metric would not only facilitate a more comprehensive understanding of each family's robustness but also aid in identifying optimal logic families for specific applications. Furthermore, it opens up the possibility of strategically mixing gates from different logic families to optimize for power, area, or latency, which can significantly enhance the overall performance and efficiency of neuromorphic systems. Consequently, we propose two distinct metrics: one that evaluates the overall resilience of a logic family and another that focuses on the resilience of individual gates within a family.

Definition 4.2 The *Quality of Logic (QoL)* is defined for a single fault model as

$$\text{QoL} = \sum_{i=0}^{G-1} \frac{\Lambda}{\Omega} \cdot 100\%, \tag{4.7}$$

where G represents the number of gate types, Λ denotes the total number of faulty outputs, and Ω signifies the total number of outputs. The QoL metric provides an indication of how well the entire set of supported logic gates performs under the influence of a specific fault type.

Definition 4.3 The *Impact of Fault (IoF)* is defined for a single gate type as

$$\text{IoF} = \sum_{i=0}^{F-1} \frac{\Lambda}{\Omega} \cdot 100\%, \tag{4.8}$$

where F is the number of fault types, Λ denotes the total number of faulty outputs, and Ω signifies the total number of outputs. This metric is particularly insightful as it reveals the impact of all fault types on a single logic gate's functionality.

By applying these metrics, we can systematically evaluate and compare the fault resilience of different logic families and individual gates, providing a comprehensive understanding of their robustness and reliability in real-world applications. This approach not only enhances our understanding of the vulnerability of these systems to in-field faults but also guides the design and optimization of neuromorphic computing systems for enhanced performance and reliability.

4.7 Evaluation

In this section, we present a detailed evaluation of in-field faults on LIM architectures, while also discussing the trade-off between simulation accuracy and speed through a comparative analysis of X-Fault and FLIM . All experiments were carried out on a dedicated workstation equipped with an AMD Ryzen 7 5800X processor and a DDR4 2666 MHz 64 GB main memory. To accelerate the simulation process, FLIM leverages the computational power of an NVIDIA GeForce RTX 3080 Ti with 12 GB of memory. Both simulators, X-Fault and FLIM , are based on a modified version of Larq 0.12.0 which utilizes TensorFlow 2.8.0 for their operations.

In the following, we delve into two case studies aimed at assessing the resilience of the LIM paradigm at varying levels of abstraction. Initially, the robustness of different logic families is exhibited through a series of comprehensive simulations using X-Fault. This is followed by an investigation of the reliability of neuromorphic applications accelerated by LIM, taking into account a variety of parameters including diverse fault models, types of layers, and BNN models.

4.7.1 Case Study: Resilience of Logic Families

To assess the resilience of various logic families, X-Fault's crossbar model is utilized to conduct an exhaustive analysis of both MAGIC and IMPLY logic families, along with their respective logic gates. The output Z of a defective logic gate depends on the input values $(x, y) = \{(x, y) \mid x, y \in \{1, 0\}\}$, the chosen fault model $f \in \{SA, RDF, DRDF, IRF\}$,

	SAF	RDF	DRDF	IRF	IoF
AND	42%	42%	39%	42%	38%
IMP	38%	25%	50%	50%	34%
NAND	33%	25%	58%	42%	34%
NOR	42%	42%	33%	42%	37%
NOT	50%	50%	75%	50%	48%
OR	33%	25%	42%	42%	31%
XNOR	44%	38%	44%	50%	41%
QoL	40%	35%	49%	45%	

	SAF	RDF	DRDF	IRF	IoF
AND	35%	25%	35%	43%	31%
NIMP	33%	29%	67%	38%	34%
NAND	45%	28%	60%	60%	45%
NOR	44%	38%	33%	42%	36%
XOR	50%	75%	50%	38%	46%
OR	34%	27%	40%	40%	30%
XNOR	44%	38%	44%	50%	40%
QoL	41%	37%	47%	44%	

(a)   (b)

Fig. 4.6 Fault resilience of logic families: (**a**) assessment of the IMPLY and (**b**) MAGIC

the number of memristors $M \in \mathbb{N}$, and the fault location $l = \{m_n \mid n \in \mathbb{N}, n \leq M\}$. To evaluate resilience, we simulated each logic gate under various fault models, considering all possible combinations of input values, initial states of memristive devices, and fault locations. We then compared the output to that of a fault-free gate to determine the percentage of incorrect outputs. The proportion of faulty outputs for all logic gates in both the MAGIC and IMPLY families is depicted in Fig. 4.6. It's important to note that coupling faults were not included in this series of experiments, as they necessitate multiple consecutive accesses. Additionally, the corresponding QoL and IoF metrics are calculated and presented in Fig. 4.6.

Noteworthy, the OR gate shows the highest resilience in both families, with an IoF of 31% for IMPLY and 30% for MAGIC. The QoL metric suggests that RDF has the least influence on logic gates, whereas DRDF significantly impacts them. Moreover, the NOT gate in the IMPLY family and the XOR gate in the MAGIC family rank as the least resilient in terms of fault tolerance. Overall, our experiments indicate comparable performance between the two families, as reflected in the similar QoL and IoF values. Nevertheless, architectural design decisions can be optimized by considering resilience towards specific fault models. For example, a designer might prefer the IMPLY implementation of the NAND gate, with an IoF of 34%, over the MAGIC version, which has an IoF of 45%.

4.7.2 Case Study: Resilience of Binary Neural Networks (BNNs)

Considering the previous findings in Sect. 4.7.1, it is crucial to gain an understanding about the impact of faults on the application level to be able to assess the reliability of the LIM paradigm. Given LIM's binary nature, BNNs are an ideal application to be accelerated with LIM operations, particularly due to their reliance on XNOR operations for inference computation. To simulate comprehensive BNN models, this case study employs the FLIM simulator. As FLIM prioritizes simulation speed at the expense of simulation accuracy, our experiments are restricted to the injection of stuck-at and bit-flip faults. To demonstrate the trade-off between X-Fault and FLIM , we benchmark both simulators, presenting a comparison of their respective performance levels. Furthermore, this study explores the impact of faults on two distinct types of layers (dense and convolutional) and their positions within the network. Finally, a range of BNN models are simulated to assess the influence of different model designs on fault impact.

4.7.2.1 Performance Evaluation

To evaluate the simulation performance of our fault injection platform, we run the inference of a pretrained binarized LeNet model on the entire MNIST test datatest composed of 10,000 images. While LeNet [13] is a convolutional neural network with three convolutional layers and two dense layers, MNIST [5] resembles a large database of handwritten digits in 28×28 pixel image. FLIM and the vanilla Larq implementation were each run fifty times on the full dataset, while the total runtime for X-Fault was extrapolated based on the analysis of just five images due to the extensive simulation runtime. During these inferences, the fault injection mechanism mapped the respective operations on a 40×10 crossbar array but did not actively inject faults, positioning the vanilla Larq implementation as a baseline for comparison in terms of simulation time.

As depicted in Fig. 4.7, FLIM significantly outperforms X-Fault. *Notably, FLIM processes the 10,000 images approximately 29,375 times faster than X-Fault.* Furthermore, leveraging the benefits of GPU acceleration through deep integration with Larq and TensorFlow, FLIM achieves a staggering *speed-up of about 66,754 times relative to X Fault.* In summary, by abstracting the fault model to the XNOR operation level, FLIM strikingly balances simulation accuracy with a remarkable boost in performance.

Fig. 4.7 Performance evaluation of the fault injection platform: running a pretrained binarized LeNet model on the MNIST dataset with FLIM and vanilla Larq, and X-Fault

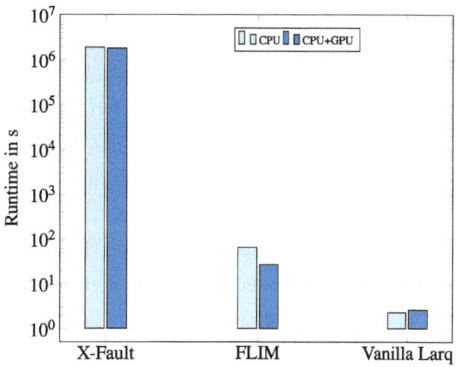

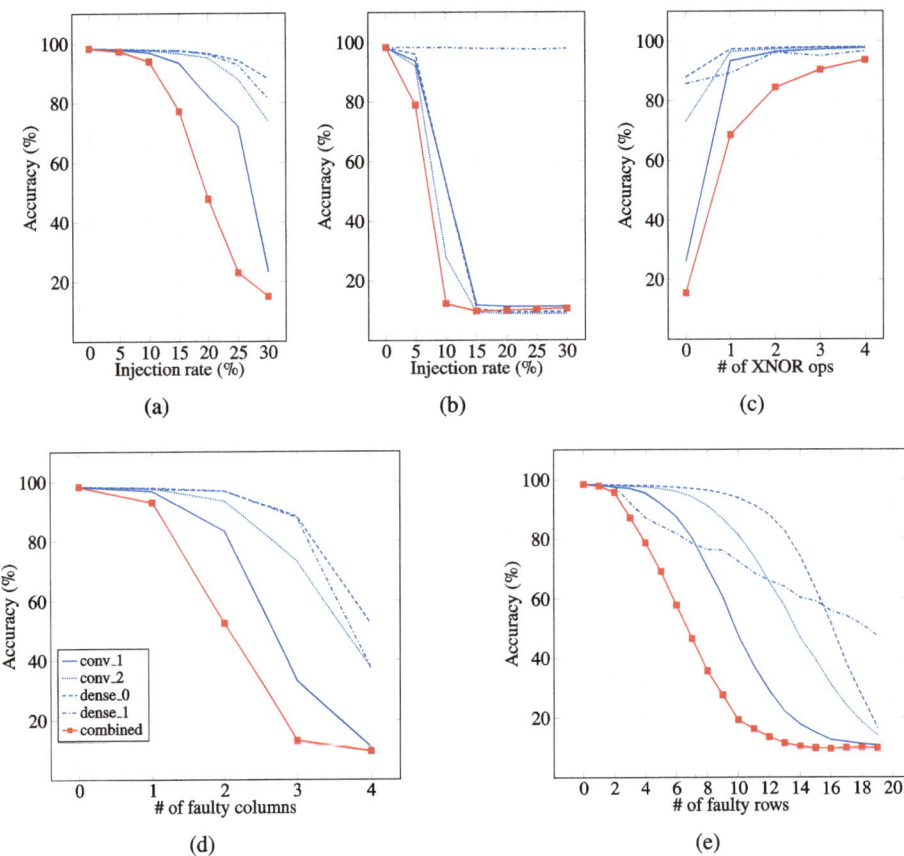

Fig. 4.8 Simulation results: impact of (**a**) bit-flips, (**b**) stuck-at, (**c**) dynamic faults, (**d**) faulty columns, and (**e**) faulty rows on different layers

4.7.2.2 Layer Resilience

This experiment is designed to explore the effects of various faults on different layers and their locations within a BNN model. Utilizing the same pretrained binarized LeNet model as in the previous Sect. 4.7.2.1, which achieves an accuracy of 97.62%, we employ the MNIST test dataset comprising 10,000 images of 28 × 28 pixels. Each layer is allocated to a 40 × 10 crossbar array for this purpose. Subsequently, we inject bit-flip, stuck-at, and dynamic faults into different crossbars corresponding to their respective layers, while adjusting the fault injection rate. To address the randomness of fault placement on the crossbars, we conducted each test run a hundred times.

The simulation outcomes are depicted in Fig. 4.8, where individual layer traces are illustrated in blue, and the red trace represents the overall accuracy impact when faults are uniformly injected across all layers. *Figure 4.8a–b demonstrate that stuck-at faults significantly degrade accuracy more than bit-flip faults, irrespective of layer type. Stuck-at faults consistently affect all layers, whereas bit-flip faults' impact on accuracy depends*

Table 4.3 Summary of BNN models and their associated parameters [8]

Model	Top-1 Acc.	Size	Parameters	MACs	Binarized
RealToBinaryNet [15]	65.0%	5.13MB	12M	1.81B	92.39%
BinaryDenseNet45 [2]	65.0%	7.54MB	13.9M	6.67B	96.34%
BinaryDenseNet37 [2]	62.9%	5.25MB	8.7M	4.71B	96.76%
BinaryDenseNet28 [2]	60.9%	4.12MB	5.13M	3.79B	94.66%
BinaryResNetE18 [11]	58.3%	4.03MB	11.7M	1.81B	92.4%
BinaryAlexNet [12]	36.3%	7.49MB	61.8M	841M	91.34%
MeliusNet22Z[3]	62.9%	3.88MB	6.94M	4.76B	97.14%
Bi-Real Net [14]	57.5%	4.03MB	11.7M	1.81B	92.4%
XNORNet [16]	45.0%	22.81MB	62.4M	1.14B	90.05%

on the layer's depth, with convolutional layers being more susceptible than dense layers. Figure 4.8c examines dynamic bit-flip faults, indicating the number of XNOR operations required for fault activation. *This analysis shows that the BNN model's accuracy typically stabilizes back to its original value after around four consecutive XNOR operations.*

Moreover, the data suggest that the layer's depth has a direct correlation with its impact on accuracy, especially highlighting a nearly linear reduction in the performance of the final dense layer, as illustrated in Fig. 4.8d–e. *Overall, faulty columns tend to have a more significant impact than faulty rows, which seems plausible due to the column-wise parallel execution of XNOR operations.*

4.7.2.3 Model Resilience

Table 4.3 lists the BNN models used in this study to examine the influence of the BNN model architecture on fault resilience. These models have been pretrained using the ImageNet dataset [6], into which both (dynamic) bit-flips and stuck-at faults were subsequently injected. To mitigate the effects of randomness from the random number generator, each inference run was repeated a hundred times.

The previous experiment established that stuck-at faults significantly reduce the accuracy more than bit-flip faults. *Figure 4.9a–b proves this finding across different BNN models, demonstrating that the impact is consistent independent of the model. Furthermore, the majority of models return to their initial accuracy levels after about three consecutive XNOR operations when dynamic faults are introduced as illustrated in Fig. 4.9c.*

Overall, it becomes evident that time-dependent fault variations influence the reliability of neuromorphic applications in diverse ways. Depending on the fault injection rate, transient faults impact the applications' reliability to different degrees. Likewise, the findings suggest that the durability of these emerging applications is predominantly

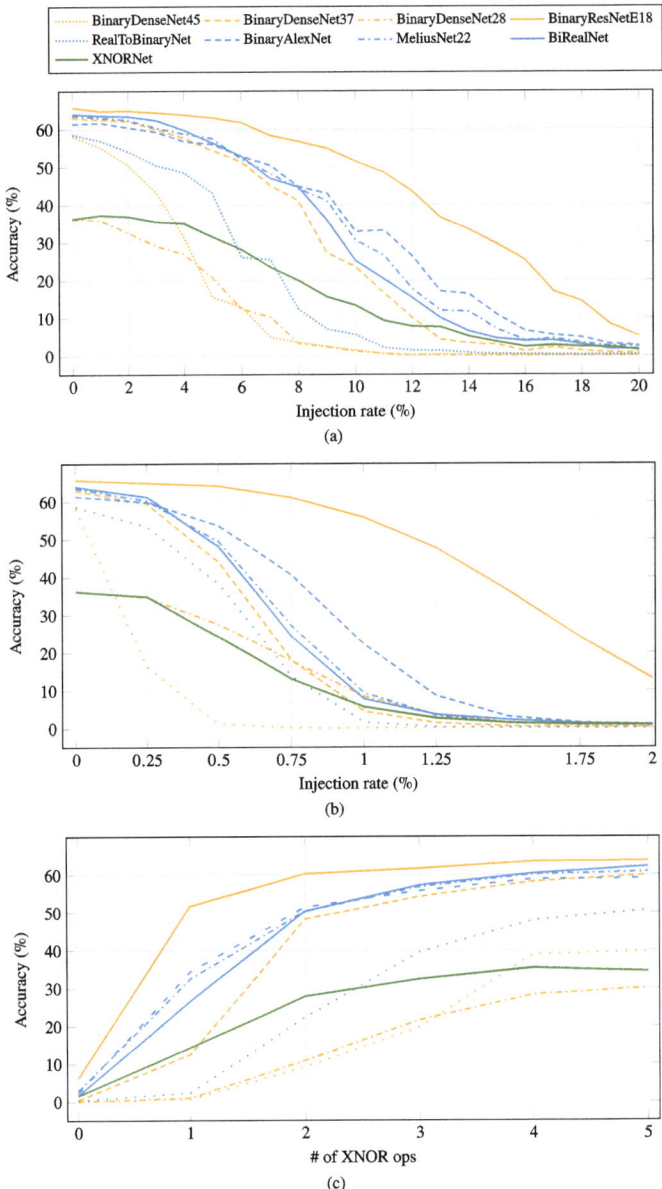

Fig. 4.9 Simulation results of (**a**) bit-flips, (**b**) stuck-at, and (**c**) dynamic faults on different models

jeopardized by permanent faults, like stuck-at faults. BiRealNet and XNOR-Net present unique cases due to their non-strict binarization approach in convolutions. BiRealNet employs real-valued activation functions through identity shortcuts [14], while XNOR-Net applies a channel-wise gain to the weights, reflecting each channel's magnitude. Despite these differences, FLIM successfully simulates both models with minor modifications to the bit-flip mask.

4.8   Limitations and Outlook

The fault injection framework introduced in this chapter has proven to be an instrumental tool for assessing the resilience of the LIM paradigm, spanning from the logic level to the application-level. The utilization of both X-Fault and FLIM offers a comprehensive strategy, assisting designers in selecting suitable logic families and validating their choices through the execution of full-fledged applications.

Nevertheless, X-Fault's performance bottleneck restricts its capacity to explore the effects of faults across extensive sets of operations. As a result, simulating only a single layer may be practical with X-Fault to keep simulation times within acceptable limits. Likewise, FLIM delivers the necessary performance for fault injection in realistic BNN models but falls short in terms of the supported fault models. Additionally, X-Fault's design limits it to logic families that rely exclusively on memristive crossbar arrays without additional peripheral circuitry. Given that most logic families require some form of external peripherals, this limitation limits X-Fault's compatibility with a wider array of logic families. In terms of FLIM , addressing the effects of in-field faults during inference is essential for understanding the practical viability of the LIM paradigm. Considering the challenges posed by the non-ideal behaviors of memristive devices, on-device training emerges as a promising solution. Therefore, expanding FLIM to include fault injection capabilities during the training phase would significantly broaden the range of applications for this fault injection framework, offering deeper insights and more robust solutions for the implementation of the LIM paradigm.

4.9   Synopsis

This chapter introduced a fault injection framework for assessing the resilience of LIM architectures. The framework consists of two simulation tools, X-Fault for memristor-level emulation and FLIM for high-speed operational simulation, to explore fault resilience across logic families and neuromorphic applications. Figure 4.1 illustrates the interaction between the two simulation platforms. X-Fault is used to assess the fault distribution, while FLIM evaluates its impact at the application level. Through comprehensive case studies, the framework evaluates the impact of faults on various logic gates and BNN models, highlighting the trade-offs between simulation accuracy and speed. A novel metric is proposed for quantifying the fault resilience of logic families and gates, aiming to enhance the understanding of their robustness against in-field faults. The chapter identifies limitations, such as X-Fault's performance bottleneck and FLIM's restricted fault model support, and suggests future enhancements, including extending FLIM to support fault injection during training. This chapter significantly advances the understanding of LIM's viability and guides the optimization of neuromorphic systems for improved reliability and efficiency.

References

1. Abadi, M., Agarwal, A., Barham, P., Brevdo, E., Chen, Z., Citro, C., Corrado, G.S., Davis, A., Dean, J., Devin, M., Ghemawat, S., Goodfellow, I., Harp, A., Irving, G., Isard, M., Jia, Y., Jozefowicz, R., Kaiser, L., Kudlur, M., Levenberg, J., Mané, D., Monga, R., Moore, S., Murray, D., Olah, C., Schuster, M., Shlens, J., Steiner, B., Sutskever, I., Talwar, K., Tucker, P., Vanhoucke, V., Vasudevan, V., Viégas, F., Vinyals, O., Warden, P., Wattenberg, M., Wicke, M., Yu, Y., Zheng, X.: TensorFlow: large-scale machine learning on heterogeneous systems (2015). Software available from https://www.tensorflow.org/
2. Bethge, J., Yang, H., Bornstein, M., Meinel, C.: Back to simplicity: how to train accurate BNNs from scratch? Preprint (2019). arXiv:1906.08637
3. Bethge, J., Bartz, C., Yang, H., Chen, Y., Meinel, C.: MeliusNet: an improved network architecture for binary neural networks. In: Conference on Applications of Computer Vision, pp. 1439–1448 (2021)
4. Chollet, F., et al.: Keras (2015). https://keras.io
5. Deng, L.: The MNIST database of HANDwritten digit images for machine learning research. IEEE Signal Process. Mag. **29**(6), 141–142 (2012). https://doi.org/10.1109/msp.2012.2211477
6. Deng, J., Dong, W., Socher, R., Li, L.J., Li, K., Fei-Fei, L.: Imagenet: a large-scale hierarchical image database. In: Conference on Computer Vision and Pattern Recognition. IEEE (2009). https://doi.org/10.1109/cvpr.2009.5206848
7. Dumoulin, V., Visin, F.: A guide to convolution arithmetic for deep learning. Preprint (2016). arXiv:1603.07285. https://doi.org/10.48550/ARXIV.1603.07285
8. Geiger, L., Team, P.: Larq: an open-source library for training binarized neural networks. J. Open Source Software **5**(45), 1746 (2020). https://doi.org/10.21105/joss.01746
9. Google Inc.: FlatBuffers (2024). https://flatbuffers.dev. Accessed Jan 24, 2024
10. Hamdioui, S., Al-Ars, Z., van de Goor, A.J., Rodgers, M.: Dynamic faults in RANDom-access-memories: concept, fault models and tests. J. Electron. Test. **19**(2), 195–205 (2003). https://doi.org/10.1023/a:1022802010738
11. He, K., Zhang, X., Ren, S., Sun, J.: Deep residual learning for image recognition. In: Conference on Computer Vision and Pattern Recognition, pp. 770–778 (2016)
12. Krizhevsky, A., Sutskever, I., Hinton, G.E.: ImageNet classification with deep convolutional neural networks. Commun. ACM **60**(6), 84–90 (2017). https://doi.org/10.1145/3065386
13. Le Cun, Y., Boser, B., Denker, J.S., Henderson, D., Howard, R.E., Hubbard, W., Jackel, L.D.: HANDwritten digit recognition with a back-propagation network. In: International Conference on Neural Information Processing Systems, NIPS'89, pp. 396–404. MIT Press, Cambridge (1989)
14. Liu, Z., Luo, W., Wu, B., Yang, X., Liu, W., Cheng, K.T.: Bi-Real net: binarizing deep network towards real-network performance. Int. J. Comput. Vision **128**(1), 202–219 (2020)

15. Martinez, B., Yang, J., Bulat, A., Tzimiropoulos, G.: Training binary neural networks with real-to-binary convolutions. Preprint (2020). arXiv:2003.11535
16. Rastegari, M., Ordonez, V., Redmon, J., Farhadi, A.: XNOR-Net: imagenet classification using binary convolutional neural networks. In: European Conference On Computer Vision, pp. 525–542. Springer (2016)
17. Staudigl, F., Sturm, K.J.X., Bartel, M., Fetz, T., Sisejkovic, D., Joseph, J.M., Pohls, L.B., Leupers, R.: X-fault: impact of faults on binary neural networks in memristor-crossbar arrays with logic-in-memory computation. In: International Conference on Artificial Intelligence Circuits and Systems (AICAS). IEEE (2022). https://doi.org/10.1109/aicas54282.2022.9869897
18. Staudigl, F., Fetz, T., Pelke, R., Sisejkovic, D., Joseph, J.M., Pöhls, L.B., Leupers, R.: Fault injection in native logic-in-memory computation on neuromorphic hardware. In: Design Automation Conference (DAC). IEEE (2023). https://doi.org/10.1109/dac56929.2023.10247742
19. Staudigl, F., Fetz, T., Pelke, R., Sisejkovic, D., Joseph, J.M., Pöhls, L.B., Leupers, R.: Invited paper: a holistic fault injection platform for neuromorphic hardware. In: Latin American Test Symposium (LATS). IEEE (2023). https://doi.org/10.1109/lats58125.2023.10154482

Deliberately Flipping Bits in Memristive Crossbar Arrays

5

Hardware security represents a crucial aspect of modern computing systems, often over-shadowed by the emphasis on software vulnerabilities. Unlike their software counterparts that can be mitigated through updates, hardware security flaws are embedded within the physical circuitry, making them almost impossible to correct without replacing the chip entirely. This intrinsic challenge poses a profound threat to the integrity and reliability of computing systems, as seen by the Meltdown [15] and Spectre [11] attacks.

In Chap. 4, we have highlighted the susceptibility of neuromorphic computing, especially the LIM paradigm, to faults that drastically undermine the reliability of applications running on neuromorphic hardware. This chapter builds upon these results by introducing a novel hardware security attack termed NeuroHammer which threatens the integrity of the entire system. This attack specifically targets the memristive crossbar arrays utilized in neuromorphic computing systems, intentionally inducing bit-flip faults to undermine the foundational principle of modern computing systems–memory separation. By exploiting the distinct properties of memristive devices to alter the switching kinetics through thermal crosstalk, NeuroHammer unveils an attack surface, similar to the Rowhammer attack in DRAMs (see Sect. 2.4.3). This chapter begins by outlining the fundamental attack scenario and the working principles underlying NeuroHammer in Sect. 5.1. To assess the feasibility and scope of the NeuroHammer attack, Sects. 5.2 and 5.3 present a simulation methodology comprising both a thermal and a circuit simulation. This dual-simulation approach enables a realistic emulation of thermal crosstalk within memristive crossbar arrays, providing insights into the conditions under which the NeuroHammer attack can be most effectively executed. Section 5.4 discusses a comprehensive set of experiments aim to evaluate the impact of various parameters on the feasibility of the NeuroHammer attack. Through these experiments, we aim to determine the critical factors that influence the vulnerability of neuromorphic computing systems to this novel hardware security threat.

© The Author(s), under exclusive license to Springer Nature Switzerland AG 2026   69
F. Staudigl, R. Leupers, *Towards Trustworthy Neuromorphic Computing*,
Synthesis Lectures on Engineering, Science, and Technology,
https://doi.org/10.1007/978-3-032-09586-2_5

To underline the real-world implications of NeuroHammer, Sect. 5.5 provides a case study which showcases the leakage of an Rivest–Shamir–Adleman (RSA) key from a computing system utilizing memristive memory. This case study underscores the seriousness of this security threat and its capacity to compromise integrity of sensitive information processed by neuromorphic computing systems. The chapter concludes with a discussion on NeuroHammer's limitations and future directions in Sect. 5.6. This chapter summarizes the contributions presented in [19, 20].

5.1   NeuroHammer

The existence of the NeuroHammer attack in passive crossbar arrays stems from two fundamental observations. Firstly, Von Witzleben et al. [24] investigate the impact of elevated temperatures on the switching kinetics of Resistive Random-Access Memories (ReRAMs), with a particular emphasis on Valence Change Material (VCM) that utilize a transition metal oxide. By employing a nanometer-scale heating structure, the authors were able to achieve a rapid temperature increase within the cell and performed kinetic measurements at different temperatures to observe the effect on the switching times and pre-SET slope. The study revealed that higher temperatures significantly decrease the SET times which also aligns with simulations results based on an analytical model.

Secondly, as discussed in Sect. 2.1.2, passive crossbar arrays necessitate some form of isolation mechanism to safeguard half-selected cells from unintended switching. The $V/2$ scheme, the most commonly adopted technique, applies a voltage drop of V_{write} across the designated device, while the non-selected devices along the word and bit lines experience an absolute voltage of $|V_{\text{write}}/2|$. Within the context of NeuroHammer, this scheme ensures that in passive crossbars, adjacent cells are exposed to a $|V_{\text{write}}/2|$ pulse with each write operation targeting the selected cell.

Merging these observations, Fig. 5.1 showcases the intrinsic workings of the Neuro-Hammer attack, represented through four discrete phases. The attack and targeted cells are

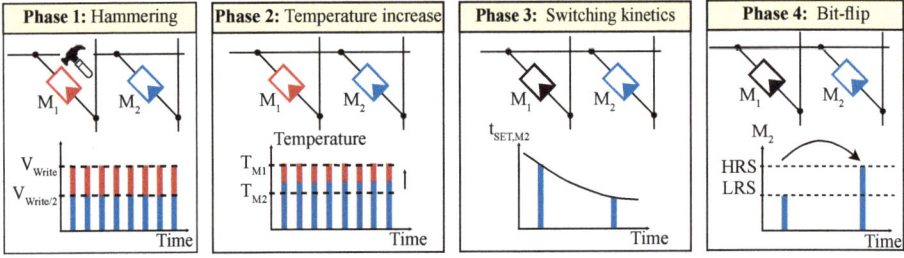

Fig. 5.1 Working principles of NeuroHammer consisting of four stages: hammering, temperature increase, switching kinetics, and, finally, the intended bit-flip

represented by red and blue cells, respectively. In the following is a comprehensive and detailed account of each phase:

1. **Hammering:** Ideally, the red cell is initially in a Low Resistive State (LRS) to maximize current flow. The attacker repetitively writes to this cell, inducing a V_{write} pulse across its terminals. Owing to the $V/2$ scheme, the blue cell endures repeated stress from the generated $V/2$ scheme.
2. **Temperature increase:** The continuous V_{write} pulses lead to a temporary rise in the red cell's temperature, which in turn increases the temperature of the adjacent blue cell due to thermal crosstalk. Concurrently, the blue cell experiences regular $|V_{write}/2|$ voltage pulses, further increasing its temperature.
3. **Switching kinetics:** As demonstrated by Von Witzleben et al. [24], increased temperatures alter the switching kinetics, reducing the SET time. Consequently, the device is more susceptible to gradually change its resistance.
4. **Bit-flip:** Eventually, the blue cell alters its internal state after a gradual change over time. The combination of thermal crosstalk and the $V/2$ scheme enables the attacker to flip a bit without direct access to the targeted cell.

This attack procedure requires a $|V_{write}/2|$ across the target cell's terminals to eventually provoke a bit-flip. However, the $V/2$ scheme introduces sneak-path currents, which constrain the crossbar size, lower the read accuracy, and increase power dissipation (see Sect. 2.1.2). Hence, 1-Transistor 1-Resistor (1T1R) structures emerge as a promising solution, leveraging transistors to insulate half-selected cells from the read/write pulses.

In the following, we validate the existence of the NeuroHammer attack on both passive and active crossbar arrays by utilizing our simulation methodology to assess the impact of thermal crosstalk. Initially, a thermal simulation confirms that thermal crosstalk sufficiently increases the temperature of adjacent cells. Additionally, the simulator yields thermal coupling coefficients, termed alpha values, for emulating crosstalk at the circuit level. A subsequent circuit-level simulation employs these alpha values to simulate thermal crosstalk, considering the electrical properties of the memristive devices and the crossbar structure. This simulation methodology not only provides key insights about the influence of thermal crosstalk in dense passive/active crossbar structures but also manifests the impact of the NeuroHammer attack in neuromorphic computing.

5.2 Thermal Simulation

In this section, we describe the implementation of a memristive crossbar model within the COMSOL Multiphysics® [6] simulation platform, aimed at quantifying thermal crosstalk between adjacent memristive cells. The simulation leverages the finite element method to determine heat transfer coefficients, referred to as *alpha values*, which are subsequently used as inputs for the circuit-level simulation (refer to Sect. 5.3). Section 5.2.1 details the

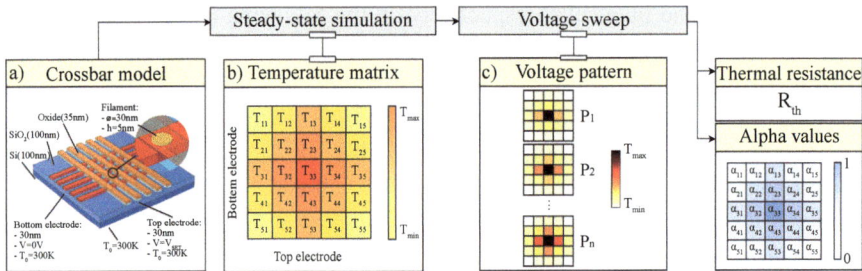

Fig. 5.2 Overview of the thermal simulation methodology: (**a**) depiction of the crossbar model alongside its boundary conditions, (**b**) extraction of device temperatures within the crossbar array, and (**c**) assessment of thermal resistance for a centered cell and the corresponding alpha values for the adjacent devices

procedural steps undertaken to evaluate thermal crosstalk originating from an individual device. Moreover, Sect. 5.2.3 elaborates on extending this methodology to analyze more advanced patterns involving multiple devices.

5.2.1   Memristive Crossbar Model

Figure 5.2a depicts the crossbar model employed for thermal simulations. This model consists of Bottom Electrode (BE) and Top Electrode (TE), which intersect perpendicularly on a Si/SiO$_2$ substrate. Unlike the JART VCM v1b compact model [1] used in circuit simulations (refer to Sect. 5.3), the memristive devices placed between the TE and BE in this model are represented solely by their filament. This approach omits the division into a resistive switching disc area and a highly conductive plug region, which significantly enhances simulation performance.

The resistance of each memristive device is thus characterized by the filament's resistance, which can be adjusted as needed. The temperature of the ReRAM cells is calculated by solving the static heat transfer equation

$$-\nabla \cdot (\kappa \nabla T) = \mathbf{j} \cdot \mathbf{E}, \tag{5.1}$$

and the current continuity equation

$$\nabla \cdot \mathbf{j} = -\nabla \cdot (\sigma \nabla \phi) = 0, \tag{5.2}$$

where T represents the temperature of the memory cell, ϕ the electrical potential, κ the thermal conductivity, σ the electrical conductivity, $\mathbf{j}$ the local current density, and $\mathbf{E}$ the electric field.

In this setup, the memristive device does not simulate any switching characteristics, meaning the heat generated by a device is determined purely by its dissipated power.

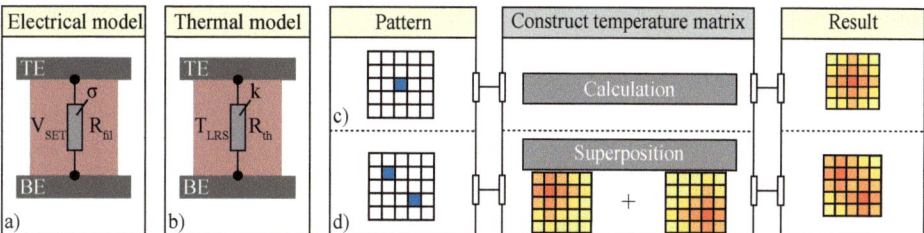

Fig. 5.3 (**a**) Electrical circuit diagram representing the modeled memristive device within the crossbar model and (**b**) its corresponding equivalent thermal diagram. (**c**) The temperature matrix featuring a single selected cell can be directly calculated. (**d**) For configurations with two selected cells, the resultant temperature matrix emerges from the superposition of two shifted temperature matrices

Consequently, the electrical conductivity of the filament is assumed to remain constant throughout the simulation. To establish a specific current I through the cell, we manually adjust the filamentary resistor's electrical conductivity R_{fil} in accordance with Ohm's law $V_{SET} = R_{fil} I$. Figure 5.3a–b illustrates the electrical and thermal model of the memristive cell, respectively.

The thermal boundary conditions are defined by the top surface of the substrate and the crossbar structure, both acting as thermal insulators. Additionally, the contacts of the electrodes and the simulation model's bottom surface are set to the ambient temperature T_0, serving as an ideal heat sink. Electrical currents are directed into or out of the memristive array exclusively via the top and bottom electrodes, with the electrode potentials set to V, $V/2$, or $0\,\mathrm{V}$.

5.2.2   Thermal Crosstalk–Single Device

Initially, we assess the thermal crosstalk generated by a single memristive cell. To isolate the effect, we assume that only the targeted cell is in an LRS, while all other cells remain in a HRS, thus minimizing the influence of self-heating from adjacent half-selected cells. Our investigation centers on the thermal crosstalk that occurs during the write process of a cell in the LRS, which maximizes the current through the targeted device and consequently, the resultant thermal crosstalk. In the context of a passive memristive crossbar array, we employ the $V/2$ scheme to selectively address the targeted cell.

Figure 5.2b presents the temperature matrix derived from a steady-state simulation of the specified crossbar model. Each element T_{ij} within the temperature matrix indicates the peak temperature of the filament at its respective position in the array. To obtain the necessary alpha values and the associated thermal resistance R_{th} of the targeted cell, we execute a voltage sweep of V_{SET}, yielding various temperature matrices. Based on Fourier's law, the temperature rise across a thermal resistor directly correlates with the heat generated from the dissipated power. Hence, we can calculate the thermal resistance R_{th}

through linear regression between the dissipated power $P_{LRS} = V_{SET}I$ and the temperature $T_{LRS}(P_{LRS})$ of the targeted cell in LRS as follows:

$$T_{LRS}(P_{LRS}) = T_0 + R_{th} \cdot P_{LRS}. \tag{5.3}$$

To derive the alpha values, we apply the same approach but note that the temperature increase dT_{ij} in an adjacent cell, due to thermal crosstalk, constitutes only a portion of the total temperature rise $dT_{LRS} = R_{th} \cdot P_{LRS}$ in the heated device. The alpha values quantify this proportion and are defined as:

$$T_{ij}(P_{LRS}) = T_0 + R_{th} \cdot P_{LRS} \cdot \alpha_{ij}, \tag{5.4}$$

where α_{ij} denotes the alpha value for a particular memristive cell located between the BE i and the TE j. The alpha value for the central cell, origin of the thermal crosstalk, is set to 1 to ensure Eq. (5.4) equals with Eq. (5.3). Therefore, alpha values for all neighboring cells will be less than 1. These values illustrate the extent of the thermal crosstalk and are influenced by the crossbar array's configuration, including electrode spacing and material properties. After determining the alpha values for a specific crossbar array, Eq. (5.4) facilitates the computation of the resultant temperature matrix for varying power dissipations of a given targeted cell.

5.2.3 Thermal Crosstalk–Multiple Devices

In this section, we develop a methodology to quantify thermal crosstalk within memristive crossbar arrays when multiple memory cells are accessed simultaneously. This approach is particularly beneficial for examining the effects of various attack patterns on the efficiency of NeuroHammer attacks. To estimate the temperature rise due to thermal crosstalk from multiple cells, we adopt the premise that heat contributions from different sources can be superimposed. As established in Sect. 5.2.2, the temperature increase of a specific memory cell is expressed as

$$dT_{ij} = R_{th} \cdot P_{ij}, \tag{5.5}$$

relying solely on the thermal resistance R_{th} and power dissipation P_{ij}. Assuming a constant thermal resistance for all cells, an adjacent device at position kl undergoes a temperature increase from the heat source at ij as described by:

$$dT_{kl}^{ij} = \alpha_{k-i+r,l-j+s} \cdot dT_{ij} \tag{5.6}$$

where r and s represent the distance from the selected cell to the top left corner of the alpha matrix. For example, considering the dimensions of a 5×5 crossbar as shown in

Fig. 5.2a, the offsets would be $r = 3$ and $s = 3$. Equation (5.6) illustrates that thermal crosstalk on an adjacent cell depends only on the relative distance and direction to the heat source, not on the absolute position of the selected cell within the array. Hence, for devices ij not positioned centrally, shifting the alpha matrix enables easy computation of thermal crosstalk. Consequently, the size of the alpha matrix does not need to correspond to the size of the crossbar array, considering that all memory cells outside the alpha matrix can be marked negligible in terms of their impact on thermal crosstalk. The cumulative temperature increase in a cell from the thermal crosstalk of all adjacent cells within the alpha matrix is given as

$$dT_{kl} = \sum_{\substack{0<i<=m \\ 0<j<=n}} dT_{kl}^{ij} \tag{5.7}$$

where m and n represent the number of rows and columns in the memristive crossbar array, respectively. The dT_{kl} considers the thermal crosstalk from all adjacent cells, including the cell's own self-heating ($k = i \wedge l = j$). The overall temperature for each device within the superimposed temperature matrix is defined as

$$T_{kl} = T_0 + dT_{kl} \tag{5.8}$$

$$= T_0 + \sum_{\substack{0<i<m \\ 0<j<n}} \alpha_{k-i+r,l-j+s} \cdot dT_{ij} \tag{5.9}$$

$$= T_0 + \sum_{\substack{0<i<m \\ 0<j<n}} \alpha_{k-i+r,l-j+s} \cdot R_{th} \cdot P_{ij}. \tag{5.10}$$

Figure 5.3c–d illustrates the methods for determining thermal crosstalk in a crossbar array. Equation (5.4) calculates the temperature of adjacent cells for a single centrally selected cell, while Eq. (5.8) is utilized for computing resultant temperatures when two or more cells are selected. Here, the thermal crosstalk contribution for each selected cell is calculated individually, followed by the superposition of these distinct temperature matrices to result in the temperature profile of the crossbar array.

5.3 Circuit Simulation

In this section, we introduce a circuit-level simulation framework designed to investigate the effectiveness and feasibility of NeuroHammer. Cadence Virtuoso® [4], serving as the base circuit simulator, incorporates our modules implemented in VerilogAMS to enable scalable simulations of crossbar structures considering thermal crosstalk. The platform is designed to be customizable via the standard graphical user interface of Virtuoso, enabling the exploration of diverse crossbar structures and experimental configurations. Figure 5.4

Fig. 5.4 Overview of the circuit simulation methodology, comprising the memory controller, the crosstalk hub, and the crossbar array

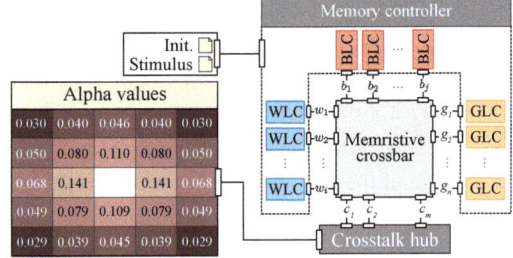

presents an overview of the simulation framework, which includes the memory controller, the crosstalk hub, and the memristive crossbar array, with a detailed discussion in the following.

5.3.1 Memory Controller

The memory controller represents the centerpiece of the circuit simulator responsible for orchestrating the essential signals for interfacing the crossbar model as shown in Fig. 5.4. Specifically, the memory controller comprises three sub-controllers, which can be instantiated based on the crossbar array's type and dimensions. The World Line Controller (WLC) and the Bit Line Controller (BLC) manage the application of appropriate read-/write/compute voltages to the targeted memory cells, with i and j denoting the number of rows and columns in the crossbar array, respectively. Similarly, the Gate Line Controller (GLC) activates the access transistors to select the memristive devices. Depending on the crossbar's architecture, n may correspond to the number of columns/rows as detailed in Sect. 2.1.2, or it may adjust to more complex configurations such as 2T1R crossbar arrays [26]. To execute various attack patterns on the crossbar, the sub-controllers must produce synchronized signals with exact timing specifications. These attack patterns are specified in a stimulus file that describes the pulse details (including length and amplitude), mapping (which allocates the pulses to specific lines), and timing. This setup is adjustable within the Cadence Virtuoso tool, allowing for parametric sweep analysis to investigate corner cases. Furthermore, an initialization (init) file specifies the initial state of each ReRAM cell.

5.3.2 Crosstalk Hub

The crosstalk hub acts as a centralized module for calculating thermal crosstalk within the crossbar array. Thus, as illustrated in Fig. 5.4, the crosstalk hub relies on the alpha matrix to indicate the influence of surrounding cells on the temperature increase of a given cell (see Sect. 5.2.1). Figure 5.5 depicts the overall procedure for determining the absolute temperature of a cell, considering thermal crosstalk and self-heating. The temperature

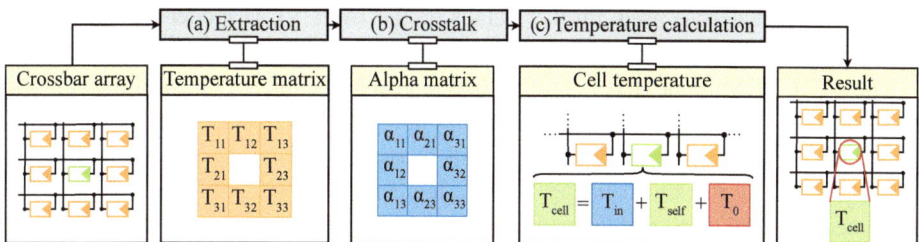

Fig. 5.5 Detailed breakdown of the crosstalk hub: (**a**) extraction of temperatures from adjacent cells, (**b**) integration with alpha matrix, and (**c**) calculation of the final cell temperature

increase due to thermal crosstalk of a given cell T_{in} can be expressed as

$$T_{in}(\alpha, \mathbf{T}) = \sum_{\substack{0<i<m \\ 0<j<n}} \alpha_{ij} T_{ij}, \tag{5.11}$$

where α represents the alpha matrix, $\mathbf{T}$ denotes the temperatures of adjacent cells, and m, n are the number of rows and columns, respectively. The crosstalk hub determines T_{in} for each cell in every simulation cycle to simulate the continuous impact of thermal crosstalk.

5.3.3   Memristive Crossbar

The memristive crossbar module represents a versatile component for creating crossbar structures. For the memristive device model, we utilize the deterministic version of the JART VCM v1b model, designed for filamentary switching in VCM cells. This model has been calibrated to a nano-crossbar Pt/HfO$_2$/TiO$_x$/Ti device [1, 7, 16].

The model internally employs the concentration of oxygen defects N_{disc} in the HfO$_2$ within a "disc" region at the Pt/HfO$_2$ interface as the state variable. These positively charged defects affect electron transport across the Pt/HfO$_2$ interface and the local conductivity. A high defect concentration signifies the device is in the LRS, while a low concentration indicates a HRS.

In general, ion migration under an applied electric field alters the defect concentration. Therefore, applying a negative voltage to the Pt electrode draws the positively charged oxygen defects, increasing N_{disc} and facilitating the SET transition. On the other hand, applying a positive voltage repels these defects, reducing N_{disc} and triggering the RESET state. The local temperature T is influenced by the dissipated power P_d according to

$$T = R_{th,eff} \cdot P_d + T_0, \tag{5.12}$$

where T_0 is the ambient temperature, and $R_{th,eff}$ represents the effective thermal resistance (in K/W), reflecting the heat dissipated to the surrounding cells and the thermal characteristics of the materials used. Further details on the JART VCM model are documented in [1], with the parameters for our experiments detailed in Appendix A.1.

The JART VCM model has been modified to enable the interaction with the crosstalk hub during simulation. Two interface variables have been introduced to relay the current device temperature to the crosstalk hub and to receive the temperature increase from adjacent cells. As illustrated in Fig. 5.5, the temperature of a specific cell T_{cell} can be computed using Eq. (5.8) and reformulated as

$$T_{cell} = T_0 + dT_{kl} \tag{5.13}$$

$$= T_0 + T_{self} + T_{in} \tag{5.14}$$

$$= T_0 + \underbrace{V_m \cdot I_m \cdot R_{th,eff}}_{T_{self}} + T_{in}, \tag{5.15}$$

where T_{cell} signifies the cell temperature, T_{in} the temperature rise due to thermal crosstalk as determined by the crosstalk hub (see Sect. 5.3.2), V_m the voltage difference across the memristor m, I_m the current through memristor m, T_0 the ambient temperature, and $R_{th,eff}$ the effective thermal resistance (in K/W).

5.4   Results

In this section, we utilize the developed simulator to explore the feasibility and effectiveness of the proposed NeuroHammer attack. Prior to delving into the specifics of the attack, we conduct a detailed analysis of the assumptions related to thermal crosstalk in memristive crossbar structures. Specifically, we validate the method of superimposing temperature matrices. Based on these findings, we validate the applicability of our simulation approach for investigating the NeuroHammer attack on 1T1R structures.

Following the thorough justification of the simulation methodology, we present a comprehensive series of experiments examining the effects of pulse length, electrode spacing, ambient temperature, and various attack patterns on the NeuroHammer attack. Additionally, we investigate different technology nodes and device variations in 1T1R structures.

5.4.1 Thermal Simulation

To understand the implications of thermal crosstalk in memristive crossbar arrays, the heat flux is studied to provide insights about the parameters influencing the strength and the propagation. Furthermore, the assumptions made in Sect. 5.2.3 have to be verified. For this experiment, we utilize the implemented crossbar array in Sect. 5.2 along the boundary conditions shown in Fig. 5.2.

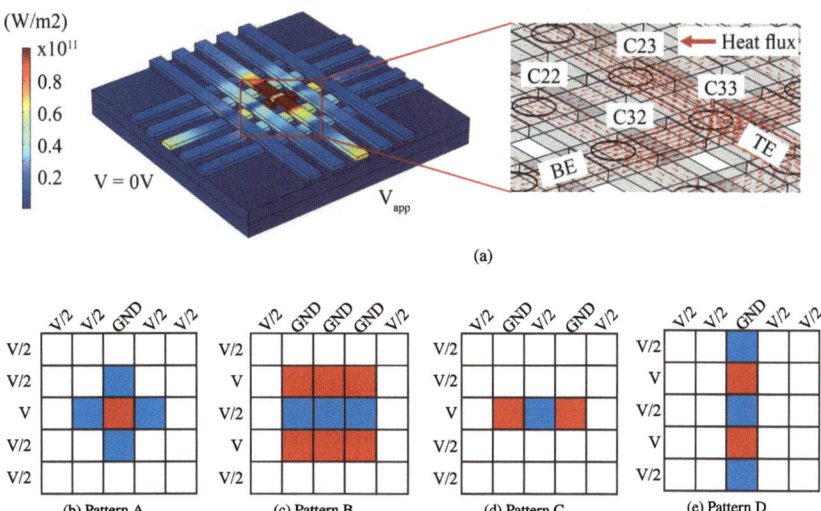

Fig. 5.6 (**a**)–(**d**) Overview of the attack patterns employed in our experiments, with blue indicating the targeted cells and red highlighting the attacked cells. (**e**) Simulation results visualizing the heat flux along the electrodes, showing that heat from cell C33 is distributed via cells C23 and C32 to the adjacent cell C22

Heat Flux Figure 5.6a visualizes the heat flux of within a passive crossbar array based on arrows. In this setup, the center cell (C33) is set to the LRS and being selected by the respective BE and TE. *The heat flux, indicated by the red arrows, spreads primarily into symmetrically into the top and bottom electrodes.* Considering that the thermal conductivity of the oxide layer under the top electrode and the substrate is significantly lower compared to that of the electrodes, the heat naturally remains in the more thermally-conductive lines. Therefore, the adjacent cells C23 and C32 experience a temperature increase due to thermal crosstalk. *Likewise, Fig. 5.6a shows that the thermal coupling decreases the farther the cells are apart from the heat source. As a consequence, the dimensions and the spacing of the crossbar array plays a crucial role for the magnitude of the thermal crosstalk.* Therefore, encasing this setup in a thermally insulating material like SiO_2 would not significantly alter the temperature profiles, since heat dispersal primarily occurs through the electrodes.

Figure 5.7a shows the alpha value of the cell (1,1) depending on the spacing between the top/bottom electrodes from one row/column to the next one. *The results illustrate a general tendency that the farther the electrodes are apart, the thermal crosstalk becomes less prevalent. However, since the selected lines also increase their temperature due to Joule heating, the alpha values may increase depending on the geometric and material properties of the electrodes.* Since the heat distributes primarily along the electrodes, the thermal crosstalk may be reduced by intentionally increasing the gap between the memristive devices and the electrodes. Such a thermal isolator may be represented by a

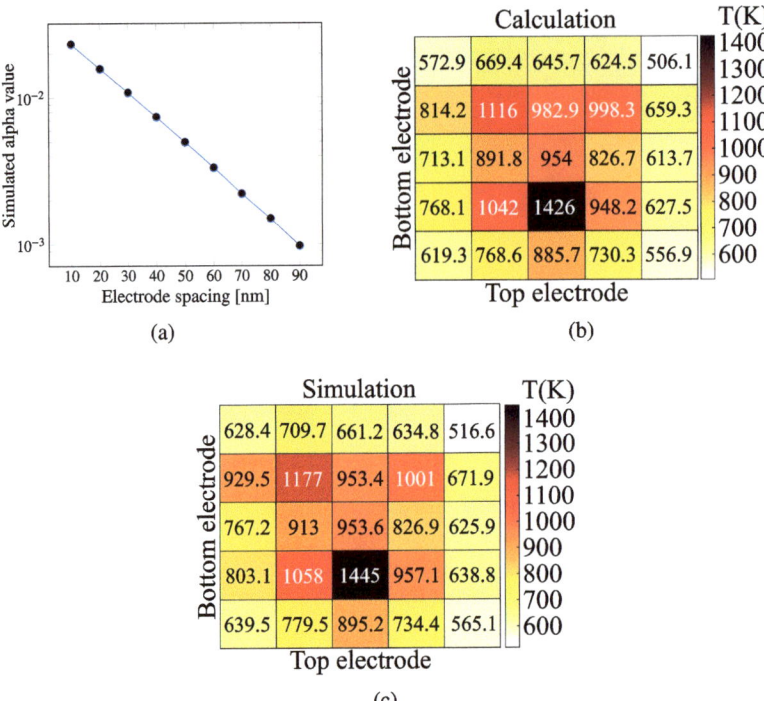

Fig. 5.7 (**a**) The alpha value of a ReRAM cell as a function of electrode spacing in a crossbar array. A comparison between (**b**) temperatures obtained from simulation and (**c**) temperatures calculated by employing the method of superimposing temperature matrices

transistor as used in 1T1R structures. *Therefore, these more complex crossbar structures can be simulated by reducing the alpha matrix to a line vector because the heat flux only propagates along the electrodes directly connected to the memristive device while the transistor effectively serves as isolator.*

Superimposing Temperature Matrices To validate the proposed method of superimposing temperature matrices, we conducted simulations on a pattern involving three selected cells in the LRS, with the remaining devices in the HRS. Figure 5.7b displays the simulated temperature matrix for this pattern, while Fig. 5.7c presents the resulting temperature matrix derived from the alpha values. *This comparison reveals that the simulated values closely match the calculated temperatures, especially for cells located in the central portion of the array.* However, discrepancies between the simulated and calculated temperatures become more noticeable for cells at the array's periphery. *In the context of investigating NeuroHammer, this level of approximation is sufficient, as the attack primarily aims to induce bit-flips in adjacent cells.* In general, the precision of our method could be enhanced by applying a larger alpha matrix to a comparatively smaller actual array size. For instance, a 9 × 9 alpha matrix could be used to approximate temperatures within a 5×5 crossbar array, ensuring comprehensive coverage of the 5×5 array, even for memory cells at the array's edges. Nevertheless, it is important to note that comparing a crossbar array to an alpha matrix of differing dimensions may introduce complexities due to variations in boundary conditions.

Von Witzleben et al. [24] conducted a study exploring the impact of temperature on the switching kinetics of memristive devices. Their findings reveal highly localized temperatures within the device, reaching approximately 1000 K, which are crucial for facilitating rapid switching behavior in VCM cells within the nanosecond timeframe. *The absolute cell temperatures produced by our simulations align closely with the measurements results on real memristive devices presented by Von Witzleben et al., thereby affirming the validity and precision of our simulation approach.*

5.4.2 1R Crossbar Arrays

In this section, we examine the NeuroHammer attack at the circuit level for passive crossbar arrays (1R), with the objective of assessing the impact of pulse length, ambient temperature, electrode spacing, and different attack patterns on the number of pulses required to induce a bit-flip. The attack patterns used in our experiments are depicted in Fig. 5.6b–e, where red tiles mark the attacked cells and blue tiles indicate the cells most susceptible to a bit-flip due to the NeuroHammer attack. The likelihood of a bit-flip in blue cells is higher because they share a common electrode with the attacked cell (as discussed in Sect. 5.4.1) and are in proximity to it. The alpha matrices utilized in these experiments are detailed in Tables A.1, A.2, A.3, A.4, A.5, A.6, A.7, A.8, A.9, A.10, and A.11.

Pulse Length This experiment explores how the length of the write pulse affects the NeuroHammer attack's efficacy. We employ a 5×5 passive crossbar with an electrode spacing of 50 nm and an ambient temperature of 300 K, using pattern A (as shown in Fig. 5.6b) to target a single cell located in the center. The memory controller's sub-controllers apply the necessary voltages to the rows and columns, with the attacked cell receiving V_{Write} and Ground (GND), while the $V/2$ scheme is used for the rest to mitigate sneak-path currents. *Figure 5.8a plots pulse lengths against the number of pulses needed for a bit-flip, showing that longer pulses reduce the required number of pulses to flip a bit in adjacent cells.*

Electrode Spacing: As shown in Sect. 5.4.1, the heat flux decreases with increasing distance to the attacked cell. To quantify this impact on the NeuroHammer attack, we investigate the influence of the electrode spacing on the NeuroHammer attack. Again, Pattern A is utilized with an ambient temperature of 300 K. Figure 5.8b illustrates the number of pulses required to trigger a bit-flip in dependence of the electrode spacing and the pulse length. *In general, the results indicate that memristive crossbars are more susceptible to NeuroHammer as the electrode spacing decreases. Consequently, we argue*

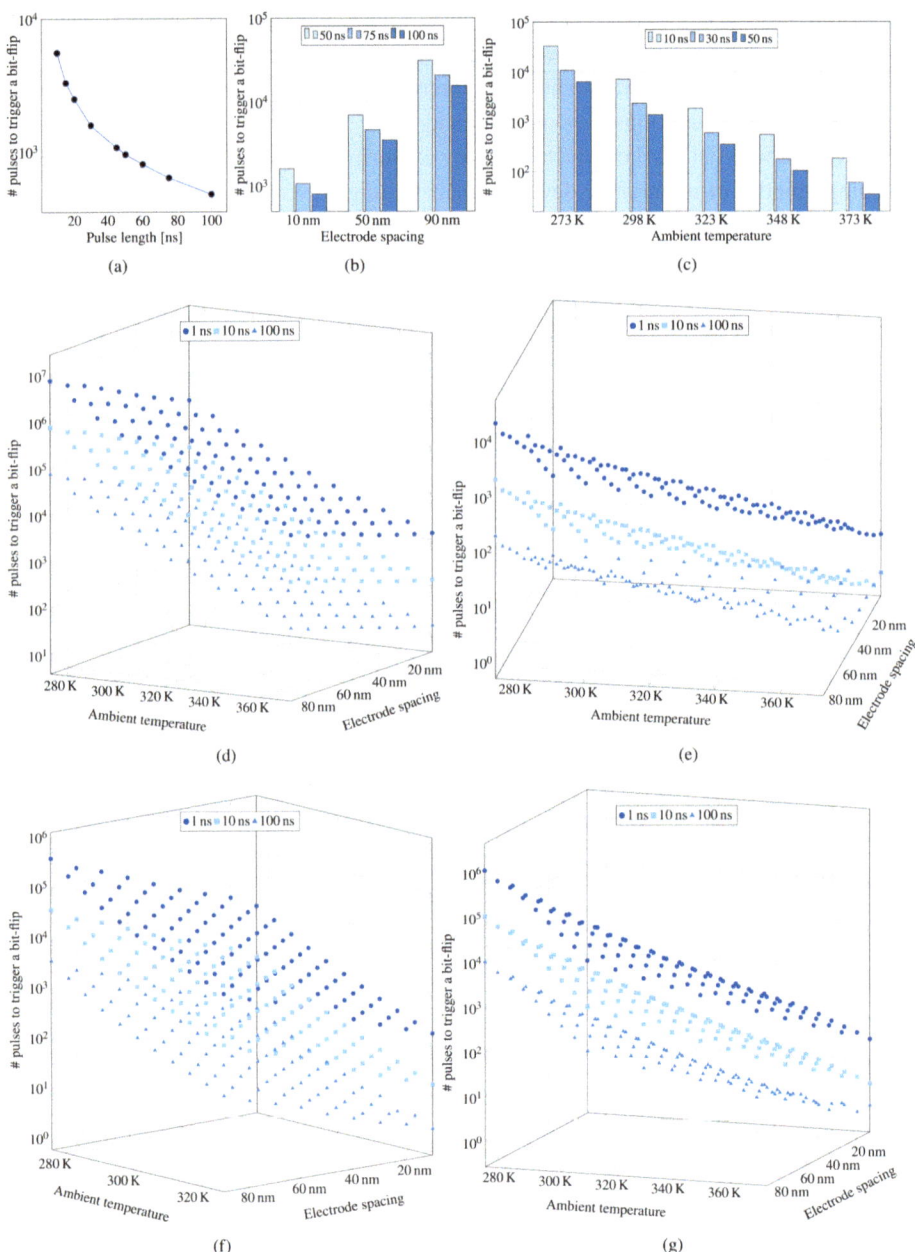

Fig. 5.8 Circuit-level simulation results for passive crossbar structures are presented as follows: (**a**) the effect of pulse length (pattern A), (**b**) the influence of electrode spacing (pattern A), (**c**) the effect of ambient temperature (pattern A), and (**d**)–(**g**) the impact of attack patterns (A–D), respectively

that security attacks like NeuroHammer become a severe problem for dense crossbar memories as the technology node advances.

Ambient Temperature   As indicated by Eq. (5.8), ambient temperature T_0 significantly influences the temperature rise in a memristive device within crossbar structures. This experiment examines the effect of ambient temperature on the number of pulses required to induce a bit-flip. We utilize attack pattern A on a passive crossbar array with an electrode spacing of 50 nm. As demonstrated in Fig. 5.8, the impact of the ambient temperature is profound. At 273 K, 33, 030 pulses are needed to cause a bit-flip, whereas at 373 K, merely 184 pulses suffice. *This finding highlights that higher ambient temperatures significantly lower the pulse count needed for a NeuroHammer-induced bit-flip, posing a potential reliability risk for memristive crossbar arrays.*

Attack Patterns   Considering that a potential adversary may not have direct access to the targeted system, it might be challenging to externally influence the parameters previously discussed. However, utilizing specific attack patterns, which involve repeatedly writing a certain bit pattern within a memory region, offers a viable strategy to enhance the NeuroHammer attack's effectiveness. Initial experiments employed the attack pattern depicted in Fig. 5.6b, targeting a single cell at the center of the crossbar array. Nonetheless, our results presented in Sect. 5.4.1 suggest that attacking multiple cells simultaneously can increase the overall heat flux within the crossbar, thereby altering the switching kinetics of the targeted cell in a way that benefits the injection of bit-flips.

Hence, this experiment explores the effects of varying attack patterns. Pattern B, see Fig. 5.6c, seeks to maximize heat flux by simultaneously attacking six cells surrounding the target cell, with the intermediary three cells showing the highest likelihood of undergoing a bit-flip due to their shared bottom electrode with the attacked cells. Pattern C, as shown in Fig. 5.6d, employs the word line as a thermal conductor connecting the attacked and target cells. The effectiveness of this pattern may depend on the memory controller's addressing scheme, potentially limiting the strategy's efficiency. Figure 5.6e presents a variation of Pattern C.

The outcome of the simulations, corresponding to Patterns A–D, are illustrated in Fig. 5.8d–g, conducted across different electrode spacings, ambient temperatures, and pulse lengths. *As anticipated, the data confirms that attacking more cells at once reduces the number of pulses needed for a bit-flip. Moreover, the proximity of the attacked cells to the target cell is critical, as heat transfer predominantly occurs through the electrodes.*

5.4.3   1T1R Crossbar Arrays

As discussed in Sect. 2.1.2, passive crossbar arrays are prone to sneak-path currents, which substantially impact their reliability. To address this, 1T1R structures have been proposed,

Table 5.1 Simulation parameters for access transistors at 180 nm and 45 nm nodes, and slow, medium, and fast memristors

Transistor	Parameter		
180nm	Name: nmos	w_{gate}: 2 μm	S/D metal width: 60 μm
	Version: v3.3	l_{gate}: 180 nm	Folding threshold: 10 μm
	Fingers: 1	V_{Gate}: 1.8 V	
45nm	Name: nmos1v	w_{gate}: 120 nm	S/D metal width: 60 μm
	Version: v6.0	l_{gate}: 45 nm	Folding threshold: 10 μm
	Fingers: 1	V_{Gate}: 1.8 V	
Memristor	Parameter		
Slow	N_{max}: 0.39×10^{26} m-3	l_{var}: 0.44 nm	V_{SET}: 1.54 V
	N_{min}: 4×10^{23} m-3	r_{var}: 40.5 nm	
Medium	N_{max}: 0.4×10^{26} m-3	l_{var}: 0.40 nm	V_{SET}: 1.54 V
	N_{min}: 8×10^{23} m-3	r_{var}: 45.0 nm	
Fast	N_{max}: 0.41×10^{26} m-3	l_{var}: 0.36 nm	V_{SET}: 1.54 V
	N_{min}: 25×10^{23} m-3	r_{var}: 49.5 nm	

incorporating an access transistor to effectively isolate unselected memory cells within the array.

Nevertheless, the NeuroHammer attack requires a trigger pulse to induce a bit-flip in the target cell. In contrast to passive crossbar arrays that deploy the $V/2$ scheme which is used as trigger pulse, such a scheme is unnecessary in 1T1R arrays. Table 2.2 provides an overview of the writing schemes utilized for the three possible 1T1R configurations– each designed to mitigate sneak-path currents, thereby countering the NeuroHammer threat. *After simulating all three 1T1R structures–typical, vertical, and pseudo–using an ideal access transistor, this assumption holds true, and no bit-flips occur.*

Consequently, the following experiments adopt the Cadence Generic Process Design Kit (GPDK) to incorporate a realistic transistor model, which facilitates the simulation of leakage currents across diverse technology nodes. Furthermore, we incorporate the model parameters described by Bengel et al. [1], which are based on measurements of memristors, each characterized by fast, medium, and slow switching behaviors. Table 5.1 shows the simulation parameters for the memristive devices and transistors. All simulations were executed at a constant temperature of 293.15 K and with a 10 nm inter-electrode gap.

Crossbars This experiment seeks to evaluate the potential for NeuroHammer attacks within 1T1R structures, utilizing a realistic transistor model for analysis. Illustrated in Fig. 5.9a–c, are the vertical, typical, and pseudo crossbar configurations, where cells under attack are marked in red and those susceptible to leakage currents are highlighted in blue. These susceptible cells are identified by the occurrence of a voltage drop across

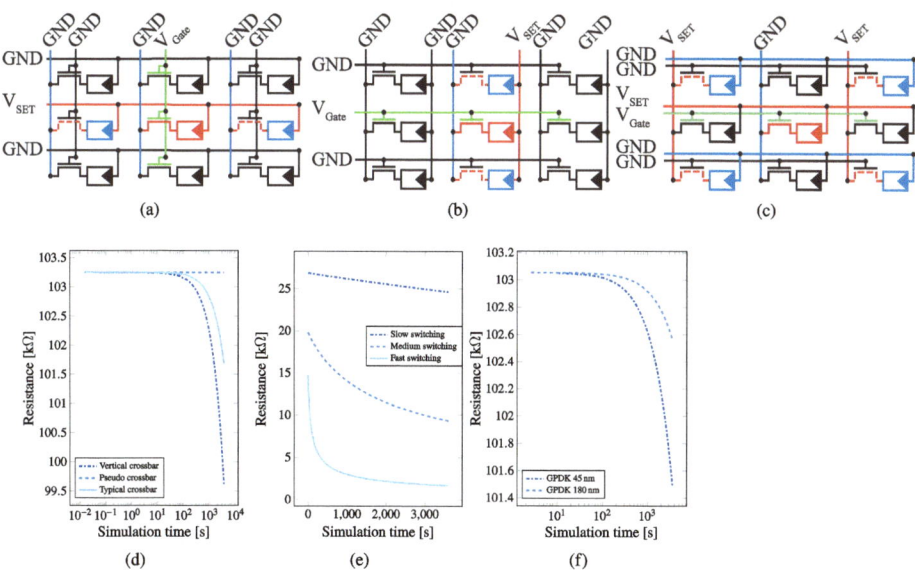

Fig. 5.9 Circuit-level simulations of (**a**) vertical, (**b**) typical, and (**c**) pseudo 1T1R crossbar structures, with impact of the crossbar structures (**d**), memristor variability (**e**), and (**f**) technology node

the memory cell which may lead to leakage currents through the access transistor. For this evaluation, we apply the fast memristor parameters together with the 45 nm access transistor. As depicted in Fig. 5.9d, we examine the resistance of the susceptible memristor throughout the simulation for each type of crossbar. *The results reveal that the leakage current passing through the access transistors could incrementally alter the resistance, potentially culminating in a bit-flip.* Findings in Sect. 5.4.1 suggest that the thermal flux predominantly traverses via the electrodes across the crossbar array. Consequently, the typical and vertical arrays demonstrate comparable susceptibility for bit-flips. *Nonetheless, the pseudo crossbar demonstrates an immunity to NeuroHammer attacks due to its unique design wherein the attacked cell and the susceptible cells do not share an electrode.* Despite experiencing similar leakage currents, the lack of thermal crosstalk within the pseudo crossbar architecture prevents substantial changes in resistance.

Memristor Variability Given that the leakage current passing through the access transistor is minimal, the intrinsic switching behavior of the memristive device becomes pivotal for the feasibility of the NeuroHammer attack. In this context, our experiment examines the influence of three distinct parameter sets for the JART VCM model, which are derived from empirical measurements of real devices. These parameters, as reported by Bengel et al. [1], correspond to fast, medium, and slow switching behaviors and are detailed in Table 5.1. The simulation results, illustrated in Fig. 5.9e, demonstrate the resistance dynamics of a vulnerable cell within a typical 1T1R crossbar structure under repeated hammering on an adjacent cell, marked in red. For this experiment, we employ attack pattern A alongside a 45 nm access transistor. It becomes evident that device variability significantly influences the success rate of the NeuroHammer attack. *The results indicate that while the memristor with slow switching attributes exhibits only marginal resistance fluctuations, the device with fast switching properties experiences a significant resistance shift, which eventually result in a bit-flip.*

Access Transistor The impact of the leakage current through the access transistor is critical in our investigation, prompting an examination of various transistor models across two distinct Generic Process Design Kit (GPDK) technology nodes. Anticipating future miniaturization of memristive crossbar arrays to enhance memory density and cost efficiency, it is projected that both memristive devices and their accompanying access transistors will experience technological scaling.

In this experiment, we utilize a typical crossbar array together with the fast memristor parameter set. *The findings, as depicted in Fig. 5.9f, confirm the hypothesis that smaller technology nodes result in higher leakage currents, which in turn, enhance the potential for a successful NeuroHammer attack.*

5.5   Case Study: Leaking RSA Keys with NeuroHammer

In Sect. 5.4, we have comprehensively investigated the existence of the NeuroHammer attack and the influence of physical parameters on its effectiveness. The ability to deliberately flip bits in memories employed in modern computing systems pose a fundamental threat as shown by Rowhammer for DRAMs (refer to Sect. 2.4.3). However, DRAM is dominantly used as main memory because of its limited access time in comparison to Static Random-Access Memory (SRAM). Therefore, the attack surface of the Rowhammer attack is bound to data stored in the main memory.

On the other hand, ReRAM , while still being heavily under research, is a promising candidate to replace not only DRAMs [13] but also SRAMs [12] for cache memories. Considering the existence of NeuroHammer, intentional bit-flip attacks are no longer restricted to main memories but may also affect cache memories. Therefore, in this case study, we explore the potential impact of NeuroHammer attacks in cache memories. In particular, we utilize the crossbar simulator presented in Sect. 4.3 and connect it to the architecture simulator Gem5 [2]. This simulation setup enables us to simulate a complete computing system including a processor, caches, and main memory.

In the following, we define a comprehensive attack scenario in Sect. 5.5.1. The implemented simulation methodology is discussed in Sect. 5.5.2 together with a thorough evaluation in Sect. 5.5.3.

5.5.1   Attack Scenario

In this case study, we assume a computing system with two users, victim and adversary, whereas the victim executes an RSA signature generation and the adversary aims to leak the secret key. Our attack scenario assumes that the attacker has the ability to execute arbitrary code within a distinct user process environment. The attacker's influence is restricted to the virtual memory space of his own process, thus prohibiting any direct modifications of data outside this context. Furthermore, the adversary also possesses thorough knowledge of the processor's architecture and its memory hierarchy. The victim uses an RSA implementation called Chinese Remainder Theorem (CRT) optimization to enhance signature generation.

Generally, RSA relies on the computational difficulty of factoring a large number N, which is the product of two large prime numbers, p and q. While an RSA signature is generated using

$$s = m^d \mod N, \tag{5.16}$$

a given message is verified by

$$s^e = m \mod N, \tag{5.17}$$

where m represents the plaintext message, $\{N, e\}$ is the public key, $\{N, d\}$ is the private key, and s is the resulting signature [8, 17]. RSA-CRT optimizes RSA by performing calculations separately for p and q, reducing the computational load by working with smaller numbers. In RSA-CRT, the signature is calculated as

$$\left.\begin{array}{l} s_1 = m^{d_p} \mod p \\ s_2 = m^{d_q} \mod q \end{array}\right\} \implies s \mod pq, \tag{5.18}$$

where $d_p = d \mod (p-1)$ and $d_q = d \mod (q-1)$. However, RSA-CRT is vulnerable to fault attacks. If an attacker induces a fault during these calculations, it can result in an incorrect signature $\tilde{s}$, computed as

$$\left.\begin{array}{l} s_1 = m^{d_p} \mod p \\ \tilde{s}_2 = m^{d_q} \mod q \end{array}\right\} \implies \tilde{s} \mod pq \tag{5.19}$$

By analyzing the difference between $\tilde{s}$ and the correct signature s, the attacker can extract the private key. Specifically, the adversary exploits the fact that $gcd(\tilde{s}^e - m, N)$ reveals p, thereby compromising the entire RSA key pair since $N = pq$ [3].

The adversary exploits this vulnerability in RSA-CRT by injecting bit-flips using NeuroHammer in the ReRAM-based L1 data cache. By disrupting the victim's signature generation process, the adversary is able to retrieve the secret key, thereby gaining full control over the signature process. Unlike traditional fault attacks, which require physical access, this scenario involves a purely software-based attack, potentially allowing for remote execution without any need for physical access to the computing system.

To the best of our knowledge, this scenario represents a novel non-intrusive attack that can actively inject faults in ReRAM caches. Furthermore, it emphasizes the profound implications of NeuroHammer on ReRAM technology and highlights the urgent need for continued research in hardware security, especially within the context of Emerging Non-Volatiles Memories (eNVMs).

5.5.2 Simulation Methodology

The Gem5 computer system architecture simulator is employed to emulate the processor, including caches and main memory. The processor model implements an out-of-order x86 processor operating at 1 GHz, equipped with L1 instruction and data caches of 16 kB each, a unified 256 kB L2 cache, and 2 GB of main memory. We have enhanced the Gem5 simulator by integrating a generic memory interface to link with the crossbar model outlined in Sect. 4.3, as depicted in Fig. 5.10. The X-Fault crossbar simulator's generic interface, illustrated in Fig. 5.11, includes a `write` and `read` function. These functions are set up as callback functions, triggered whenever the processor attempts to access cache memory. Originally, X-Fault's crossbar simulator monitored read, write, and compute accesses for each cell. However, to enhance our simulation's efficiency, we have deactivated this tracking feature and simplified the crossbar model to solely accommodate the essential fault model. The fault model is parameterized by a specific pattern and a threshold value, which dictate the number of write accesses required to trigger a bit-flip in an adjacent cell.

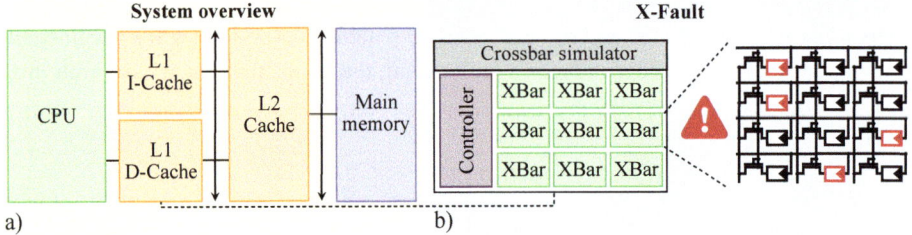

Fig. 5.10 (**a**) Overview of the simulated computing system including processor model, cache memories and main memory. (**b**) X-Fault's crossbar simulator integrated in the Gem5 architecture simulator

```
#ifndef H_FAULTYMEMIF_HPP
#define H_FAULTYMEMIF_HPP
#include <cstdint>
#define GEM5_PACKET_MAXSIZE 64 /* in bytes */
typedef enum {PAGE_INTERLEAVE, RANK_INTERLEAVE} addr_trans_t;

class FaultyMemIF {
    public:
        virtual int size_gigabytes(void) = 0;
        virtual uint64_t size_kilobytes(void) = 0;
        virtual void write(uint64_t phys_addr,
                           uint8_t *src,
                           uint_fast16_t len) = 0;
        virtual void read(uint64_t phys_addr,
                          uint8_t *dst,
                          uint_fast16_t len) = 0;
        FaultyMemIF(addr_trans_t transltr) {
            buf_len = GEM5_PACKET_MAXSIZE/sizeof(uint64_t);
            translator = transltr;
        }
        virtual ~FaultyMemIF() {};
    protected:
        uint64_t       buffer[GEM5_PACKET_MAXSIZE/sizeof(uint64_t)];
        uint_fast16_t buf_len;
        addr_trans_t   translator;
};
#endif /* H_FAULTYMEMIF_HPP */
```
Fig. 5.11 Abstract memory interface of X-Fault's crossbar simulator. The interface defines the required functions to enable the interaction between gem5 and the crossbar simulator

5.5.3   Evaluation

In this section, we detail the steps an adversary must undertake to successfully induce a bit-flip at the precise location within the L1 data cache during the victim's RSA signature generation process as shown in Fig. 5.12. For an adversary to introduce bit-flips into a specific memory cell, it is crucial to comprehensively profile the cache memory during the RSA signature generation, aiming to achieve two objectives. Initially, the attacker needs to identify the cache sets utilized during the signature process and, subsequently, select an address of a neighboring cache line that will serve as the target for the attack.

Target Cache Sets Identifying a potential target cache set begins with the adversary allocating a substantial amount of data and accessing it. This action ensures that the data populates the L1 data cache, spanning multiple cache sets. Then, by executing an RSA signature generation, the attacker profiles the cache to identify which cache sets are being evicted. By repeating this process, it becomes apparent which addresses are frequently evicted, indicating a high likelihood of containing data used during the victim's signature generation. Figure 5.13a displays a heatmap of the L1 data cache activity during signature

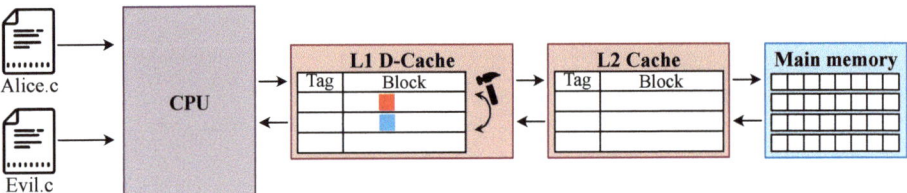

Fig. 5.12 Attack Scenario: leveraging NeuroHammer to induce bit-flips in the L1 data cache, ultimately compromising the victim's secret key

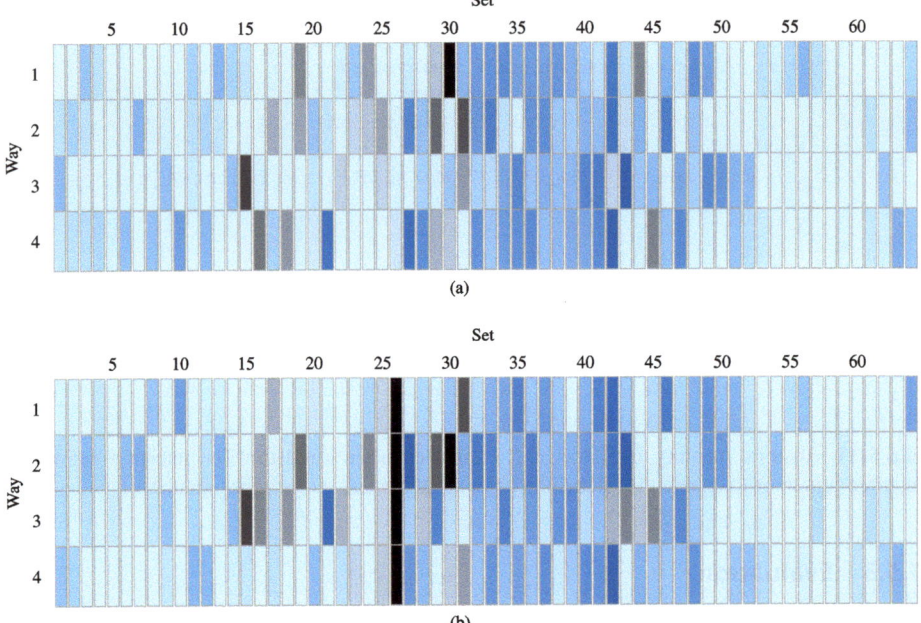

Fig. 5.13 Write access patterns to the L1 data cache during RSA signature generation: (**a**) in the absence of an attacker, and (**b**) with an active attacker targeting the 26th cache set

generation, where light blue represents minimal write activity and dark blue/black signifies intense activity. To compromise the integrity of the signature generation process, the attacker aims to inject a single bit-flip in a heavily utilized cache set, where adjacent sets are infrequently accessed. As an example, cache set 15 as depicted in Fig. 5.13a would fulfil this requirement.

Adjacent Address The method to determine an adjacent address varies based on the provided memory allocation technique. If the CPU architecture permits the use of larger-than-standard pages, known as huge pages, the attacker can easily choose a cache set by altering the lower bits of the address within the huge page. With standard page sizes, the attacker may adjust the virtual address to acquire the address of an adjacent cache

---

Algorithm 5.1: Injecting bit-flip during RSA-CRT signature generation with Neuro-Hammer.

Data: Target address (t), victim user (*victim*), message (*msg*)

1 EvSet $ev \leftarrow$ getEvSet(t-64);
2 $(e, N) \leftarrow$ getPubKey(*victim*);
3 sendSignReq(victim, msg);
4 **while** $!sig = recvSignResp(victim)$ **do**
5   **for** *Addr a* **in** *ev* **do**
6   a.data $\leftarrow$ 0b0000...1...0;
7   a.data $\leftarrow$ 0b0000...0...0;
8   **end**
9 **end**
10 **if** *sig.valid* **then**
11   **return** 0;
12 **else**
13   **return** $gcd(\mathrm{sig}^e - \mathrm{m}, N)$;
14 **end**

---

set. Considering a cache entry of 64 B, the six Least Significant Bits (LSBs) of the virtual address determine the offset. The following bits of the virtual address specify the corresponding cache set. By inverting the seventh bit of the virtual address, which represents the LSB of the set index, an address adjacent to the original address in the cache is generated.

L1 Cache Attack Once the adjacent memory cell's address has been identified, the attacker proceeds with the actual attack by repeatedly hammering the targeted cell. The pseudocode for this method is outlined in Algorithm 5.1. This algorithm requires the target address, which correlates with a cache set active during RSA signature generation, the victim user to initiate the signature process, and an arbitrary message. To enhance the attack's efficacy, we leverage eviction sets to target the entire cache set rather than a single entry. Eviction sets consist of a collection of virtual addresses that all map to the same cache set [22]. Constructing eviction sets can be achieved through either a top-down approach [18] or a bottom-up strategy [21], making the attack agnostic to the cache's replacement policy. Attacking a single address risks the RSA-CRT data being located in a different cache way than the intended target. By employing eviction sets, faults are induced across all adjacent memory cells linked to the eviction set's addresses, potentially causing unintended application crashes, although this was not observed in our experiments. Additionally, the flipped bits are unlikely to be propagated to the main memory since the entries are not marked as *dirty*. Next, the attacker requests a signature from the victim while simultaneously hammering the adjacent cache line. Figure 5.13b illustrates the L1 data cache's simulated write activity during the attack, with cache set 26 distinctly as showing intense activity marked in black, aiming to induce bit-flips in the adjacent 27th set. Our simulations consistently managed to inject bit-flips within the RSA-CRT signature generation using the NeuroHammer attack. As elaborated in Sect. 5.5.1, injecting a fault during the computing of s_1 and s_2 results in a faulty signature $\tilde{s}$. Finally, the adversary can retrieve the private key of the victim by calculating $p = \gcd(\tilde{s}^e - m, N)$.

5.5.4 Additional Attack Targets

This section provides a brief overview of potential targets susceptible to the NeuroHammer attack, extending the scenario illustrated in Fig. 5.12. It assumes that both L1 and Last Level Cache (LLC) are composed of memristive crossbar arrays, rendering them vulnerable to NeuroHammer.

Last Level Cache (LLC) Attack In our initial analysis, we considered a situation where both the victim and the attacker share a single processor core via hyper-threading, allowing access to the same L1 cache. However, in systems prioritizing security, this feature might be deactivated, restricting attackers from accessing the shared L1 cache memory. Under these circumstances, an adversary might target the LLC for the attack. For an *inclusive* LLC, the attack procedure is relatively straightforward: all write operations to the L1 cache are also immediately reflected in the LLC. Thus, any bit-flips induced in the L1 cache will similarly affect the LLC. However, it's unlikely that the bit-flip impacts the identical data in both caches, a nuance the attacker must account for. On the other hand, *non inclusive* LLCs inherently block access to other cores' private caches, thereby hindering NeuroHammer's ability to induce bit-flips directly [5]. Consequently, the adversary needs to identify a cache set that contains data critical for the victim's signature generation process yet lies outside the victim's core private cache. While locating such a cache set poses a significant challenge, the attacker benefits from essentially limitless attempts, with a single successful bit-flip revealing the secret key.

Tag/Flag Bit Attacks The NeuroHammer attack allows an attacker to arbitrarily flip bits in memristive memories. Thus, a potential adversary might aim at the tag and flag bits within a cache entry, leading to data corruption and possibly data loss. To manipulate the tag section of an adjacent cache set, an adversary must frequently access an eviction set that exceeds the cache's associativity. This action results in the tag section being continuously updated with various tags, eventually triggering a bit-flip fault in the targeted entry. Such an occurrence could cause incorrect cache hits for certain addresses or even intentionally alter the tag of a dirty entry to disrupt write operations. In addition to targeting the tag section, an attacker might also focus on the flag section of a cache entry. For example, by altering the valid flag, the adversary can falsely mark specific cache entries as valid, leading to data corruption and inconsistent cache states.

5.6 Limitations and Outlook

Given the impact of the NeuroHammer attack identified in our study, it's important to discuss the limitations of our investigation. Although our dual-simulation approach–combining detailed thermal simulation with higher-level circuit analysis–demonstrates considerable accuracy, the actual verification of thermal crosstalk and the existence of the NeuroHammer attack on real hardware remains unexplored. Our simulation, inspired by crossbar arrays fabricated in various research labs [9, 10, 14, 23, 25], may not fully capture the complexities of thermal crosstalk in fully integrated systems. Moreover, more sophisticated crossbar structures, such as the 2T1R configuration, may be potentially immune to the NeuroHammer attack which our study does not address. Additionally, the attack scenario inspired by the attack model of Rowhammer, may not cover the entire range of possible real-world threats, possibly overlooking capabilities and constraints of an attacker.

To bridge these gaps, a more holistic simulation approach is required by considering a wider range of crossbar structures and calibrate our simulation with measurements of physical structures. Future work should aim to identify new potential attack vectors, extend the attack scenario to cover a broader spectrum of threats, and crucially, investigate effective countermeasures and mitigation techniques.

5.7 Synopsis

This chapter presents NeuroHammer, a novel hardware security threat that compromises memory integrity by causing bit-flips in memristive crossbar arrays. We have adopted a dual-simulation approach to examine thermal crosstalk in nanoscale crossbar structures, which represents the fundamental mechanism behind NeuroHammer. Our extensive analysis of system parameters confirms the attack's feasibility, demonstrating its effectiveness and limitations. Through a case study, we highlight the practical consequences of NeuroHammer by detailing an attack scenario where cache memories are targeted. The chapter concludes by outlining further research perspectives to refine our simulation approach and validate it against actual crossbar array hardware.

References

1. Bengel, C., Siemon, A., Cuppers, F., Hoffmann-Eifert, S., Hardtdegen, A., von Witzleben, M., Hellmich, L., Waser, R., Menzel, S.: Variability-aware modeling of filamentary oxide-based bipolar resistive switching cells using SPICE level compact models. Trans. Circuits Syst. I Regular Papers **67**(12), 4618–4630 (2020). https://doi.org/10.1109/tcsi.2020.3018502
2. Binkert, N., Beckmann, B., Black, G., Reinhardt, S.K., Saidi, A., Basu, A., Hestness, J., Hower, D.R., Krishna, T., Sardashti, S., Sen, R., Sewell, K., Shoaib, M., Vaish, N., Hill, M.D., Wood, D.A.: The Gem5 simulator. ACM SIGARCH Comput. Archit. News **39**(2), 1–7 (2011). https://doi.org/10.1145/2024716.2024718

3. Boneh, D., DeMillo, R.A., Lipton, R.J.: On the importance of eliminating errors in cryptographic computations. J. Cryptol. **14**(2), 101–119 (2000). https://doi.org/10.1007/s001450010016

4. Cadence Design Systems Inc.: Virtuoso System Design Platform (2024). https://www.cadence.com/en_US/home/tools/custom-ic-analog-rf-design/virtuoso-studio.html. Accessed Jan 14, 2024

5. Chaudhuri, M.: Zero inclusion victim: isolating core caches from inclusive last-level cache evictions. In: Annual International Symposium on Computer Architecture (ISCA). IEEE (2021). https://doi.org/10.1109/isca52012.2021.00015

6. COMSOL AB: COMSOL Multiphysics® (2023). https://www.comsol.com/comsol-multiphysics. Accessed Sep 14, 2023

7. Cüppers, F., Menzel, S., Bengel, C., Hardtdegen, A., von Witzleben, M., Böttger, U., Waser, R., Hoffmann-Eifert, S.: Exploiting the switching dynamics of HfO2-based ReRAM devices for reliable analog memristive behavior. APL Mater. **7**(9) (2019). https://doi.org/10.1063/1.5108654

8. Gennaro, R., Krawczyk, H., Rabin, T.: RSA-Based Undeniable Signatures, pp. 132–149. Springer Berlin Heidelberg, Berlin (1997). https://doi.org/10.1007/bfb0052232

9. Kim, H., Mahmoodi, M.R., Nili, H., Strukov, D.B.: 4k-memristor analog-grade passive crossbar circuit. Nature Commun. **12**(1) (2021). https://doi.org/10.1038/s41467-021-25455-0

10. Kim, K.H., Gaba, S., Wheeler, D., Cruz-Albrecht, J.M., Hussain, T., Srinivasa, N., Lu, W.: A functional hybrid memristor crossbar-array/cmos system for data storage and neuromorphic applications. Nano Lett. **12**(1), 389–395 (2011). https://doi.org/10.1021/nl203687n

11. Kocher, P., Horn, J., Fogh, A., Genkin, D., Gruss, D., Haas, W., Hamburg, M., Lipp, M., Mangard, S., Prescher, T., Schwarz, M., Yarom, Y.: Spectre attacks: exploiting speculative execution. In: IEEE Symposium on Security and Privacy (S&P). IEEE (2019). https://doi.org/10.1109/sp.2019.00002

12. Lee, M.J., Lee, C.B., Lee, D., Lee, S.R., Chang, M., Hur, J.H., Kim, Y.B., Kim, C.J., Seo, D.H., Seo, S., Chung, U.I., Yoo, I.K., Kim, K.: A fast, high-endurance and scalable non-volatile memory device made from asymmetric Ta2O5-X/TaO2-X bilayer structures. Nature Mater. **10**(8), 625–630 (2011). https://doi.org/10.1038/nmat3070

13. Li, H.H., Chen, Y., Liu, C., Strachan, J.P., Davila, N.: Looking ahead for resistive memory technology: a broad perspective on rera technology for future storage and computing. IEEE Consumer Electron. Mag. **6**(1), 94–103 (2017). https://doi.org/10.1109/mce.2016.2614523

14. Li, H., Wang, S., Zhang, X., Wang, W., Yang, R., Sun, Z., Feng, W., Lin, P., Wang, Z., Sun, L., Yao, Y.: Memristive crossbar arrays for storage and computing applications. Adv. Intell. Syst. **18**, 309–323 (2021). https://doi.org/10.1038/s41563-019-0291-x

15. Lipp, M., Schwarz, M., Gruss, D., Prescher, T., Haas, W., Fogh, A., Horn, J., Mangard, S., Kocher, P., Genkin, D., Yarom, Y., Hamburg, M.: Meltdown: reading kernel memory from user space. In: USENIX Security Symposium (USENIX Security 18), pp. 973–990. USENIX Association, Baltimore (2018). https://www.usenix.org/conference/usenixsecurity18/presentation/lipp

16. Menzel, S.: Juelich Aachen Resistive Switching Tools (JART) (2024). http://www.emrl.de/Jart.html. Accessed Jan 14, 2024

17. Rivest, R.L., Shamir, A., Adleman, L.M.: Cryptographic Communications System AND Method (1983). US Patent 4,405,829

18. Song, W., Liu, P.: Dynamically finding minimal eviction sets can be quicker than you think for side-channel attacks against the LLC. International Symposium on Research in Attacks, Intrusions and Defenses (RAID 2019) pp. 427–442 (2019)

19. Staudigl, F., Indari, H.A., Schon, D., Sisejkovic, D., Merchant, F., Joseph, J.M., Rana, V., Menzel, S., Leupers, R.: NeuroHammer: inducing bit-flips in memristive crossbar memories. In: Design, Automation & Test in Europe Conference & Exhibition (DATE). IEEE (2022). https://doi.org/10.23919/date54114.2022.9774651

20. Staudigl, F., Al Indari, H., Schön, D., Chen, H.Y., Sisejkovic, D., Joseph, J.M., Rana, V., Menzel, S., Hagelauer, A., Leupers, R.: It's getting hot in here: hardware security implications of thermal crosstalk on ReRAMs. IEEE Trans. Reliab. 1–15 (2024). https://doi.org/10.1109/tr.2024.3371589

21. Thoma, J.P., Güneysu, T.: Write me and i'll tell you secrets - write-after-write effects on intel CPUs. In: International Symposium on Research in Attacks, Intrusions and Defenses, RAID. ACM (2022)

22. Vila, P., Kopf, B., Morales, J.F.: Theory and practice of finding eviction sets. In: IEEE Symposium on Security AND Privacy (SP). IEEE (2019). https://doi.org/10.1109/sp.2019.00042

23. Wan, W., Kubendran, R., Schaefer, C., Eryilmaz, S.B., Zhang, W., Wu, D., Deiss, S., Raina, P., Qian, H., Gao, B., Joshi, S., Wu, H., Wong, H.S.P., Cauwenberghs, G.: A compute-in-memory chip based on resistive RANDom-access memory. Nature **608**(7923), 504–512 (2022). https://doi.org/10.1038/s41586-022-04992-8

24. von Witzleben, M., Fleck, K., Funck, C., Baumkötter, B., Zuric, M., Idt, A., Breuer, T., Waser, R., Böttger, U., Menzel, S.: Investigation of the impact of high temperatures on the switching kinetics of redox-based resistive switching cells using a highspeed nanoheater. Adv. Electron. Mater. **3**(12), 1700294 (2017)

25. Xia, Q., Yang, J.J.: Memristive crossbar arrays for brain-inspired computing. Nature Mater. **18**(4), 309–323 (2019)

26. Ye, W., Wang, L., Zhou, Z., An, J., Li, W., Gao, H., Li, Z., Yue, J., Hu, H., Xu, X., Yang, J., Liu, J., Shang, D., Zhang, F., Tian, J., Dou, C., Liu, Q., Liu, M.: A 28-nm RRAM computing-in-memory macro using weighted hybrid 2T1R cell array and reference subtracting sense amplifier for AI edge inference. IEEE J. Solid-State Circuits **58**(10), 2839–2850 (2023). https://doi.org/10.1109/jssc.2023.3280357

Instrumentation Platform for Non-Volatile Memory Technologies

The previous chapters have underscored the significant influence of memristive devices' nonidealities on the reliability and security of neuromorphic computing systems. Although the discussed simulation methodologies aim to replicate these inherent properties accurately, simulation alone cannot offer a comprehensive analysis that incorporates real memristive devices and circuits.

As highlighted in Sect. 3.4, various instrumentation platforms for emerging non-volatile memory technologies have been developed, primarily focusing on characterizing memristive devices. Yet, existing platforms fall short in performing CIM operations on memristive crossbar arrays, leading to a reliance on simulation for most reliability and hardware security research.

Therefore, this chapter introduces the NBB , a versatile and adaptable instrumentation platform designed to investigate the characteristics of memristive devices at the device, crossbar, and operational levels. Equipped with custom-designed signal generation and sensing circuitry, the NBB enables precise control over memristive cell programming and supports the execution of both analog CIM and LIM operations. Additionally, the platform features three distinct application interfaces, facilitating seamless integration of its measurement capabilities across different levels of analysis.

Sections 6.1 and 6.2 explore the hardware and software components of the NeuroBreak-outBoard, followed by a comprehensive case study to demonstrate the platform's unique functionalities in Sect. 6.3. The chapter concludes with a discussion on the platform's limitations and potential future enhancements. This chapter summarizes the contributions detailed in [19].

© The Author(s), under exclusive license to Springer Nature Switzerland AG 2026
F. Staudigl, R. Leupers, *Towards Trustworthy Neuromorphic Computing*,
Synthesis Lectures on Engineering, Science, and Technology,
https://doi.org/10.1007/978-3-032-09586-2_6

6.1   Hardware

In this section, we detail the hardware components and the circuitry implemented for the NeuroBreakoutBoard as shown in Fig. 6.1a. The signal path of the NeuroBreakoutBoard is divided into three main parts: signal generation, routing, and sensing. Additionally, the board provides two hardware interfaces. The first interface, termed the NVM interface, facilitates easy integration of a wide variety of memristive crossbar arrays. The second interface allows for the connection of a controller to the board.

6.1.1   Signal Generation

The signal generation module of the NeuroBreakoutBoard, as shown in Fig. 6.2a, comprises a DAC , an analog switch, and an amplifier circuit. The DAC is responsible for providing the voltages required for writing, reading, and computing in the crossbar array. The component is managed via a Serial Peripheral Interface (SPI) bus and features eight channels, each offering a 12-bit resolution across multiple voltage ranges: $0\,V$ to $5\,V$, $0\,V$ to $10\,V$, $\pm 5\,V$, $\pm 10\,V$, and $\pm 2.5\,V$, all powered by an internal high-precision reference [7]. Out of these, five channels are dedicated to providing the interconnection matrix (refer to Sect. 6.1.2) with the required V_{SET}, V_{RESET}, V_{Read}, V_X, and V_{Gate} voltages, while the remaining three channels are connected to a pin header for additional uses.

Given that the writing algorithms of memristive devices often rely on pulsed signals, a simple but efficient pulse generator, using an analog switch with a switching time of $t_{ON} = 60\,ns$ at $\pm 5\,V$, has been implemented [9]. This setup, comprising five analog

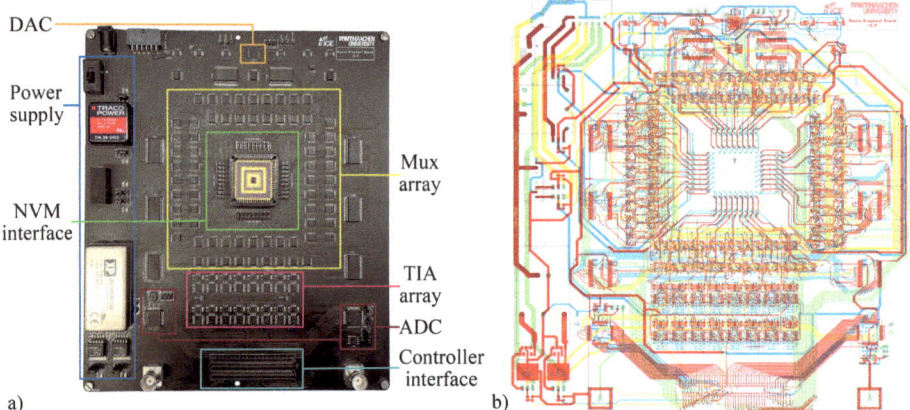

DAC

Power supply

NVM interface

Mux array

TIA array

ADC

Controller interface

a)   b)

Fig. 6.1 Overview of the NeuroBreakoutBoard: (**a**) image of the manufactured Printed Circuit Board (PCB) and (**b**) multi-layer layout

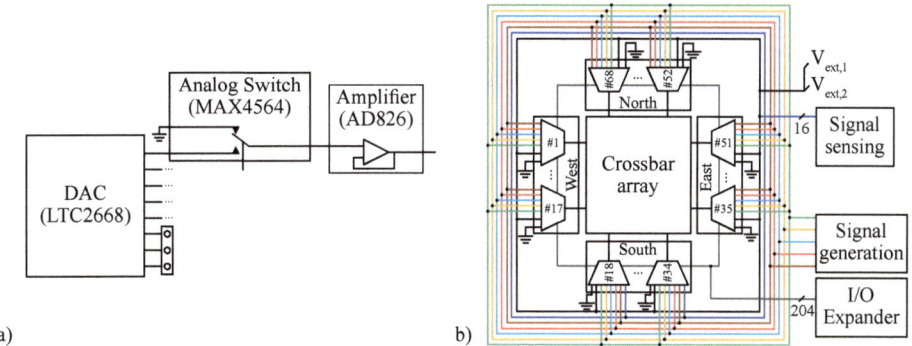

Fig. 6.2 Overview of (**a**) the signal generation module, comprising the DAC and pulse generator, and (**b**) the interconnection matrix that links the signal generation/sensing modules to the NVM interface

switches, is linked to the controller interface, enabling independent, rapid, and precise pulse generation.

While read and write operations within a crossbar array typically require minimal current, the current demand significantly increases during analog vector/matrix multiplication operations, especially as the dimensions of the crossbar array expand. To accommodate currents up to several milliamperes, which exceed the DAC's output capability, the NeuroBreakoutBoard incorporates load drivers placed between the pulse generator and the interconnection matrix. These load drivers, built from operational amplifiers, are capable of handling a maximum current of 50 mA at 15 V [2].

In summary, the signal generation module of the NeuroBreakoutBoard is designed for versatility, allowing for the adjustment of pulse amplitudes and durations. This feature not only supports standard read/write operations but also facilitates CIM tasks within memristive crossbar arrays by driving the necessary currents effectively.

6.1.2 Flexible Interconnection Matrix

The signal generation module connects to the NVM interface via a flexible interconnection matrix, enabling arbitrary mapping of input pulses to the crossbar array through the use of 68 precision analog multiplexers [3]. Figure 6.2b illustrates the architecture of the interconnection matrix, including its links to the signal generation and sensing modules. Each multiplexer can route one physical line of the crossbar to eight distinct signals. The V_{SET}, V_{RESET}, and V_{Read} signals facilitate programming and reading a memristive cell. V_X is versatile, serving either to apply $V/2$ to non-selected cells in a passive array or to set any chosen potential for computing voltage in LIM operations. Additionally, V_{Gate} is connected to the gates of access transistors in 1T1R arrays. However, these signals can be arbitrarily assigned to any potential, with the only limitation being the DAC's capabilities within the signal generation module. The $V_{ext,1}$ and $V_{ext,2}$ signals are connected to a pin header to serve as probe lines, allowing external measurement devices to connect to the crossbar array and enhance the NBB's measurement capabilities. To operate the multiplexers, each requiring three control bits for signal selection, results in a total of 204 control wires. By utilizing an I/O expander, the typically limited number of digital output signals of a microcontroller is accommodated. Multiplexers at the interconnection's south end are unique in that one line connects to the signal sensing module, allowing for 16 sensing lines to be exclusively mapped to these specific multiplexers on the crossbar interface's south side.

6.1.3   Signal Sensing

To measure the resistance of a memory cell or the result of a CIM operation, it's necessary to determine the current flowing along a bit line. Therefore, we implemented the signal sensing module which offers a broad measurement range while maintaining a high accuracy. The sensing module consists of a TIA [8] circuit connected to an ADC [1], as depicted in Fig. 6.3a. Since the ADC is limited to only convert voltages, the implemented TIA circuit translates the cumulative current from the crossbar array's bit lines into a measurable voltage. Given the varying resistances of memristive devices due to their switching processes, the TIA incorporates a dynamically adjustable feedback resistor. This feature allows for the measurement of a wide spectrum of resistances with precise accuracy. The feedback resistor of the TIA can be adjusted via an analog switch, with values ranging from 43 Ω to 100 kΩ [4]. Furthermore, an amplifier is employed to amplify the converted voltages to fit into the measurable range of the ADC . The ADC has eight parallel channels, each with an 18-bit resolution, capable of simultaneous sampling. In total, the NBB facilitates the measurement of 16 parallel channels while the results can be transmitted to the controller through either an SPI bus or a parallel interface. To increase the accuracy of the ADC , a dedicated voltage reference IC featuring a high precision of

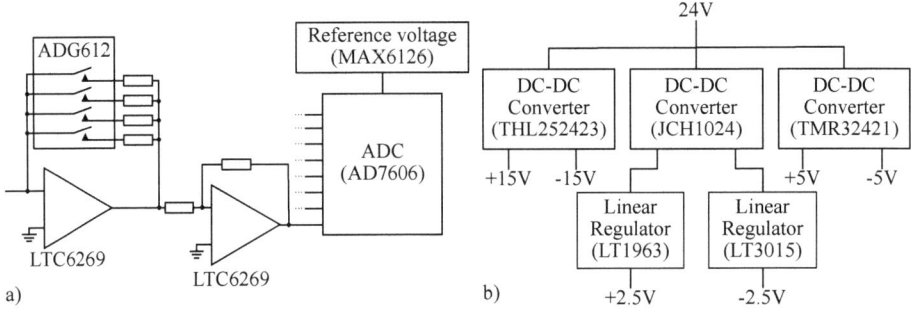

Fig. 6.3 Overview of (**a**) the signal sensing module, which includes a TIA connected to the ADC , and (**b**) the power supply module that offers three distinct power levels

up to 0.02% [10]. Additionally, an external voltage reference can be provided via a BNC connector.

6.1.4 Power Supply

To accommodate a wide range of memristive devices, the NBB is capable of supplying a wide set of voltages for both its internal modules and the crossbar array. Figure 6.3b depicts the NBB's power supply, which is designed to operate with a standard 24 V power adapter which ensures an independent usage of the measurement platform from external laboratory equipment. Nevertheless, the NBB also includes an additional socket for connecting external high-precision voltage sources. The PCB supports three distinct voltage domains: -2.5 V to 2.5 V, -5 V to 5 V, and -15 V to 15 V, each supplied by dedicated DC-DC converters, enabling a versatile power supply for various memristive device requirements [5, 6, 21–23].

6.1.5 Non-Volatile Memory (NVM) Interface

The NVM interface primarily consists of four pin headers located at the center of the NBB , as depicted in Fig. 6.1a. This interface facilitates straightforward connections to adapter boards, thus enabling the integration of additional circuitry and their corresponding sockets. Figure 6.4a presents an adapter board designed for a memristive crossbar housed in a PGA100 package. Notably, the adapter board provides extra digital I/O pins, which are instrumental in managing the digital control logic embedded in the chip under test. The adapter board's I/O expanders are linked to the NBB's controller via external wires, ensuring the synchronization of all input signals to the chip. The interconnection matrix significantly enhances flexibility in signal routing. Utilizing the NVM interface in conjunction with specially designed adapter boards allows the NBB to interface with

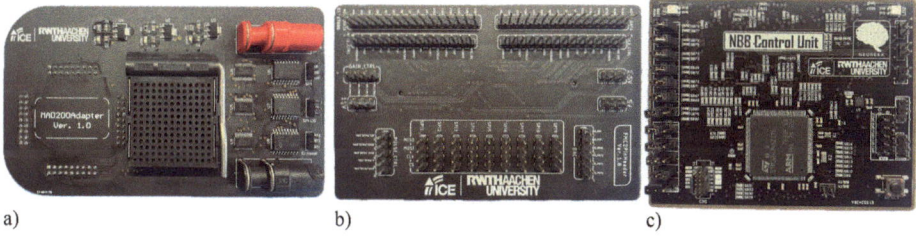

a)   b)   c)

Fig. 6.4 Extension boards: (**a**) adapter board linking to the NVM interface of the NBB , providing a dedicated socket and additional external digital I/Os, (**b**) breakout board for direct access to all control lines connected to the control interface, and (**c**) controller unit featuring a microcontroller that executes the implemented firmware

any chip package and pin configuration available. However, the application of the NVM interface is constrained by the configuration of the sensing module. Given the limited number of parallel sensing channels—specifically, 16—these independent channels are restricted to routing exclusively to the north pin header of the NVM interface. Nonetheless, the interconnection matrix provides two additional sensing wires, which can be utilized to connect with external measurement equipment.

6.1.6 Platform Orchestration

While the NVM interface serves as a universal abstraction for connecting to crossbar arrays, the controller interface acts as an abstract interface for managing the NBB platform. An FPGA Mezzanine Card (FMC) connector bundles control lines from all modules on the NBB , utilizing a standardized connector primarily employed in FPGAs to provide a vast array of I/O pins. To facilitate easy integration with various development platforms, an FMC breakout board has been developed.

This breakout board, depicted in Fig. 6.4b, allocates specific pins for each SPI interface of the I/O expanders, as well as the DAC and ADC . It also includes the ADC's parallel interface, complete with all control pins, enabling full access to the ADC's diverse functionalities. Additionally, gain control pins are linked to analog switches that adjust the feedback resistances of the TIAs, as well as the pulse control pins responsible for initiating pulses during read, write, and compute operations.

Figure 6.4c illustrates a control unit equipped with a STM32 microcontroller from STMicroelectronics [20], directly connected to the controller interface via the FMC connector at the PCB's bottom side. The microcontroller interfaces with all components through the SPI bus, while employing the ADC's parallel interface for enhanced performance during repetitive measurements. The unit is designed for flexibility, featuring pin headers and jumpers for straightforward modification of fixed configuration pins and includes status LEDs to signal the measurement process or detect errors. Moreover, the control unit is equipped with a Joint Test Action Group (JTAG)/Serial Wire Debug (SWD) connector for programming the microcontroller and a Universal Asynchronous Receiver/Transmitter (UART) bus interface for communication with the host PC.

6.2 Software

This section discusses the specifics of the software implemented on the control unit and the host PC, designed to conduct measurements and execute CIM operations on physical crossbar arrays. The architecture of the overall system is depicted in Fig. 6.5, showcasing the NeuroBreakoutBoard with the integrated control unit connected to a host PC via a UART bus. Communication between the control unit and the host PC employs the Protocol Buffers framework [14], which serializes data in a platform-independent manner. The

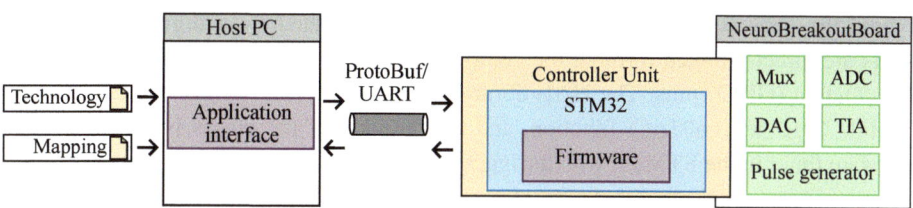

Fig. 6.5 Overview of the system architecture: the NeuroBreakoutBoard consists of hardware modules orchestrated by firmware running on the controller unit. This controller unit communicates via a UART/ProtoBuf [14] protocol with the provided application interfaces

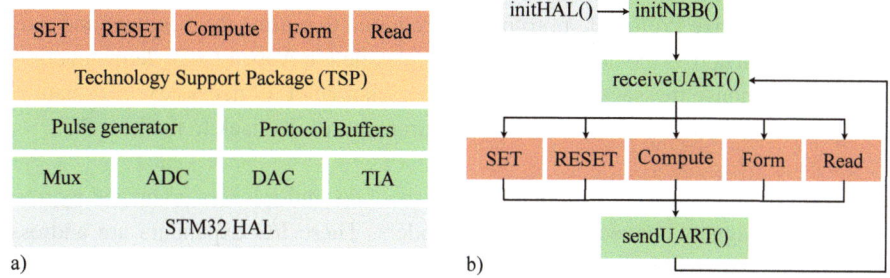

Fig. 6.6 NeuroBreakoutBoard firmware: (**a**) software modules implemented to offer various abstraction levels for interfacing with the hardware modules, and (**b**) flowchart depicting the main loop

application interface, executed on the host PC, is developed in Python and boasts a robust Application Programming Interface (API) that facilitates the execution of experiments. Additionally, it supports the incorporation of more sophisticated tools, enhancing its functionality for advanced experimental setups.

6.2.1 Firmware

The firmware architecture, as shown in Fig. 6.6a, is organized in layers and developed in ANSI-C. The STM32 Hardware Abstraction Layer (HAL) offers extensive APIs for interfacing with the microcontroller's peripherals, simplifying the development of user applications through efficient management of communication peripherals, data transfers, interrupts, Direct Memory Access (DMA), and error handling. Moreover, HAL incorporates runtime failure detection to improve firmware robustness and facilitate debugging. On top of HAL , software modules abstract the NeuroBreakoutBoard's peripherals, providing interfaces for controlling the pulse generator, multiplexers, ADCs, DACs, TIAs, and for encoding/decoding protocol buffer messages, which are elaborated in the following.

Pulse Generator The pulse generator is designed to produce the necessary pulses for read, write, and compute operations. The pulses' amplitude is adjusted by the DAC

, while the generator itself uses an analog switch activated by a digital input. These inputs are directly linked through the control interface to the STM32's General-Purpose Input/Output (GPIO) pins. The HAL library facilitates interaction with the GPIOs using the HAL_GPIO_WritePin() function, and the duration of the pulses is controlled by a hardware timer in the STM32, which triggers an interrupt event.

Protocol Buffers Protocol Buffers (ProtoBuf) is an open-source project that aims to provide a framework for serializing structured data that is both language and platform-independent [14]. Within our measurement platform, ProtoBuf is employed to serialize data for transmission over UART between the host PC's instrumentation application and the STM32. To utilize ProtoBuf, a proto file is required to outline the structure and size of the messages sent across the bus. The proto files, detailed in Appendix B, define operation, request, and response messages, facilitating the exchange of all necessary information to execute operations on the NBB and relay measurement results back to the host PC.

Multiplexer The interconnection matrix consists of 68 multiplexers, managed by a total of 204 control signals connected to I/O expanders. These I/O expanders are addressed over a SPI bus allowing a straightforward orchestration of the whole interconnection matrix. The STM32 HAL library provides the HAL_SPI_Transmit() function, enabling message transmission over the SPI bus interface. Beyond enabling the interaction with I/O expanders, the multiplexer module also introduces a mapping function. This function allows for the configuration of any pin within the interconnection matrix to connect to an output pin of the DAC , significantly simplifying algorithm implementation at a higher abstraction level.

Analog-Digital Converters (ADCs) ADCs serve as the core of the sensing module, offering both parallel and serial interfaces for data transmission to the control unit. With the SPI controllers of the STM32 microcontroller already occupied by the I/O expanders and the DAC , the ADC's parallel interface connects to GPIO pins for data transmission. The module features an adcInit() function for ADC and GPIO setup. The calculateResistance() function computes the resistance of a memristive device as follows:

$$R_{\text{cell}} = \frac{3V_{\text{Read}}}{V_{\text{ADC}}R_{\text{Feedback}}}, \tag{6.1}$$

where R_{cell} represents the memristive cell's resistance, V_{Read} the read voltage, V_{ADC} the TIA's output voltage, and R_{Feedback} the TIA's feedback resistance. Additionally, the calculateCurrent() function determines the cumulative current along a specified bit line, essential for analog CIM operation.

Digital-Analog Converters (DACs) The DAC , pivotal for generating the required pulses for reading, writing, and computing, connects to the microcontroller via the SPI bus. The DAC firmware module enables users to set the reference voltage range and modify output voltages. Given the microcontroller's limited SPI interfaces, the DAC shares an SPI module with the I/O expanders, necessitating the setting of the respective chip select pin connected to a GPIO pin on the microcontroller before message transmission.

Before setting the actual output voltage, the DAC's voltage reference must be configured. The `setSoftSpanRange()` function sends the appropriate SPI message to adjust the DAC's reference voltage. The `setupDACVoltageByChannel()` function then establishes the desired output voltage for a specific channel.

Transimpedance Amplifiers (TIAs) The transimpedance amplifier module regulates the feedback resistance, determining the sensing module's measurement resolution. The feedback resistances are connected to analog switches, managed by four digital control bits. On the NBB , these control bits for all sixteen TIAs are routed in parallel and mapped to the controller interface, setting the feedback resistance uniformly across all TIA modules. The module includes the `setTIAFeedbackResistance()` function for adjusting the feedback resistance of the TIA modules.

Technology Support Package (TSP) Following Fig. 6.6a, situated above the core firmware modules, the TSP incorporates routines specific to various technologies. Memristive devices, for example, necessitate distinct writing schemes to enhance their reliability and extend their lifespan. Consequently, the TSP is tailored for each type of memristor technology, refining the foundational access functions at this level. The TSP offers an interface comprising five functions, enabling the mapping of both memory and CIM operations on crossbar arrays. This flexibility not only broadens the measurement platform's range of applications but also allows for extensive customization to accommodate various memristive device technologies.

The primary routine of the NBB firmware is outlined in Fig. 6.6b. Upon the platform's startup, the firmware initializes the STM32 HAL library, followed by the configuration of all implemented modules. Specifically, multiplexers are set to their initial states, and both the ADC and DAC are reset. The main loop includes a blocking `receiveUART()` function call, waiting on a ProtoBuf message from the host PC (Figs. B.1, B.2, and B.3). After decoding the message's contents, the corresponding function is executed. The response message, consisting of a status flag alongside the resistances or currents measured by the sensing module, is then transmitted.

6.2.2 Application Interfaces

The application interface, designed to run on a host PC, is developed in the platform-independent programming language, Python. This interface connects to the NBB firmware

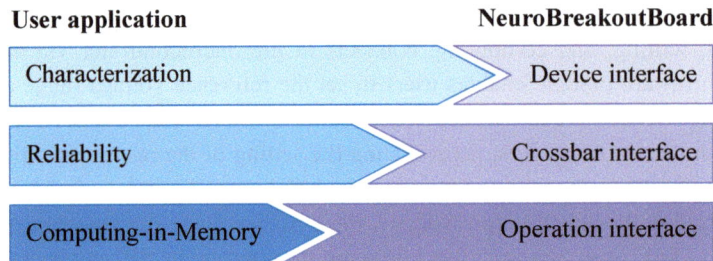

Fig. 6.7 Application interfaces: the provided application interfaces facilitate interaction with the NeuroBreakoutBoard across three distinct abstraction levels, enabling everything from device characterization to the execution of CIM operations

as depicted in Fig. 6.6a, offering a comprehensive middle layer for conducting both device/crossbar measurements and executing analog/digital CIM operations. The application interface, shown in Fig. 6.7 provides three distinct abstraction layers through which external applications can interact with the NBB , facilitating a unique method to interact with neuromorphic crossbar arrays. This approach enables early adoption and seamless integration into the software development lifecycle. The specifics of these interfaces are elaborated in the following.

Device Interface The device interface serves as the most fundamental layer, designed to facilitate direct interaction with memristive devices within a crossbar array. The formCell() function executes the forming process essential for initializing a memristive cell. Given the variety of memristive devices, this routine may be adjusted to align with the specifications of the memristive crossbar array connected to the NeuroBreakoutBoard. Currently, we have incorporated the Incremental Form and Verify (IFV) algorithm, which enhances the switching characteristics and post-forming yield, as highlighted in [15]. The IFV algorithm progressively increases the forming voltage by a given ΔV_{form}, conducting a read operation after each pulse. The process continues until the cell resistance meets a predefined target current i_{target}.

Subsequent to forming, the readCell() function facilitates cell reading, whereas SET/RESET operations are managed by the setCell() and resetCell() functions, respectively. The write functions employs the Incremental Step Pulse with Verify Algorithm (ISPVA), adopting a similar write/verify approach as the IFV algorithm. Notably, the read function leverages an automatic feedback resistance algorithm to select the TIA's feedback resistance based on the measurement outcome. This algorithm iteratively adjusts the feedback resistor, in case the operational amplifier is saturated, by choosing the next feedback resistance and recursively invoking itself.

Device addressing is managed through a configuration file that maps the physical chip pins to the corresponding pins of the NeuroBreakoutBoard's interconnection matrix, which must be manually created before employing the instrumentation application.

In addition to the provided functions, the device interface offers an advanced logging feature that records all executed commands for each device, providing a comprehensive history of the cells within the crossbar array. This data is stored in a platform-independent CSV file, encapsulating all necessary parameters to replicate operations and analyze C2C variations, D2D differences, and the yield before and after forming.

Crossbar Interface Compared to the device interface, the crossbar interface encompasses read(), set(), reset(), and form() functions that operate on an entire crossbar array. These are crucial for initializing the crossbar for various experiments, including analog/digital CIM operations. Inputs to these functions are matrices corresponding to the crossbar's dimensions, efficiently utilizing the underlying device interface for executing write and read operations. Therefore, all executed actions are automatically tracked in the provided database provided by the device interface.

Operation Interface The operation interface represents the highest level of abstraction, enabling the execution of CIM operations on a crossbar array. Given that memristive crossbar arrays can process data in both digital and analog modes, this interface is separated into two distinct modules.

For conducting analog vector/matrix multiplications within a crossbar structure, the matrix B must be encoded as conductance values across the memristive devices, with the input vector a applied as voltages across the word lines. This generates an output vector c, represented by the cumulative current along the bit lines. Assuming $a_i \in \{0, 1\}$ and $B_{i,j} \in \{0, 1\}$, inputs are converted from $\{0, 1\} \rightarrow \{0\,V, V_{read}\}$ and matrix values from $\{0, 1\} \rightarrow \{G_{min}, G_{max}\}$. The resulting vector undergoes a reverse conversion through the demap() function, which adjusts each value according to

$$c_m = \frac{1}{V_{read}(G_{max} - G_{min})}(I_{BL,m} - V_{read}G_{min}\sum_{0}^{N-1} a_{B,n}),\qquad(6.2)$$

where c_m indicates the converted result for bit line m, $I_{BL,m}$ the total current for bit line m, N the row count, and V_{read} the applied read voltage. Currently, this module supports binary vector matrix multiplication only, with further quantization methods requiring additional mapping algorithms.

Executing LIM operations necessitates selecting and implementing a suitable logic family, each significantly differing in operation. The operation interface supports this process by providing utility functions and maintaining a predefined data structure for tracking input and output devices, along with compute voltages and the physical representations of the Boolean values.

Additionally, this interface expands the device interface's tracking capabilities to encompass comprehensive logging of CIM operations. Given the potential engagement of multiple or all cells in a single operation, a specialized operation log file is created and linked to the device-specific history files. This log file records operation types, compute voltages, and result values, offering a detailed account of the experimental procedures and results.

6.3   Case Study: Reliability Assessment of a Commercial ReRAM Technology

This case study aims to showcase the versatility of the NeuroBreakoutBoard by examining the reliability of a commercially available ReRAM technology node.

We designed and taped out memristive memory composed of a 1T1R crossbar structure. The memory cell comprises a Metal-Insulator-Metal (MIM) stack, functioning as the memristive device, alongside an access transistor serving as the selection device. The MIM stack consists of top and bottom electrodes, each approximately 150 nm thick, made of TiN, a scavenging layer of about 7 nm Ti, and a dielectric switching layer of 8 nm HfO_2, resulting in a total MIM stack area of 600 nm × 600 nm. The scavenging layer undergoes oxidation during the memristive device's forming process, introducing oxygen vacancies into the MIM stack and creating a conductive filament within the dielectric layer. The conductive filament's internal structure, manipulated by applying SET or RESET voltages, determines the cell's resistance. The MIM stack is positioned between the second and third metal layers in the BEOL phase of the fabrication process. The accompanying access transistor is manufactured using 130 nm CMOS technology, featuring gate dimensions of 130 nm × 150 nm [16–18]. While the crossbar array is designed with dimensions of 12 × 7, this study focuses on a 7 × 7 submatrix. The word, bit, and gate lines are directly connected to the chip's pads and wire-bonded to a Quad Flat Package (QFP). An adapter board, tailored for the chip's socket, links directly to the NVM interface of the NeuroBreakoutBoard.

In the following, we assess the memristive crossbar structures for their ability to perform digital and analog CIM operations. Initially, we evaluate the yield before and after forming, switching endurance, and binary and multi-level switching characteristics of the fabricated memory cells. Upon investigating these properties, we explore the feasibility of executing CIM operations. This involves conducting a vector/matrix multiplication to observe the effects of faults on computational results. Finally, we investigate the crossbar array's susceptibility to faults through the execution of logic operations, further elucidating the robustness and versatility of these memristive systems.

6.3.1   Manufacturing Yield

Typically, a memristive device is anticipated to be in the HRS before the forming process. Milo et al. [16] define the manufacturing yield of memristive devices as the percentage of cells exhibiting a read current below 1 μA prior to forming. We assessed the resistance

Table 6.1 Overview of parameters for the ISPVA and the IFV algorithm

State	V_{Gate} [V]	I_{Target} [μA]	V_{Start} [V]	V_{Max} [V]	V_{Read} [V]	V_{Step} [V]	T_{Pulse} [μs]
Forming	1.50	30.00	2.00	5.00	0.20	0.01	1.00
HRS	3.30	1.00	0.50	3.00	0.20	0.10	10.00
LRS1	0.50	10.00	0.50	3.00	0.20	0.10	10.00
LRS2	0.70	20.00	0.50	3.00	0.20	0.10	10.00
LRS3	0.90	30.00	0.50	3.00	0.20	0.10	10.00

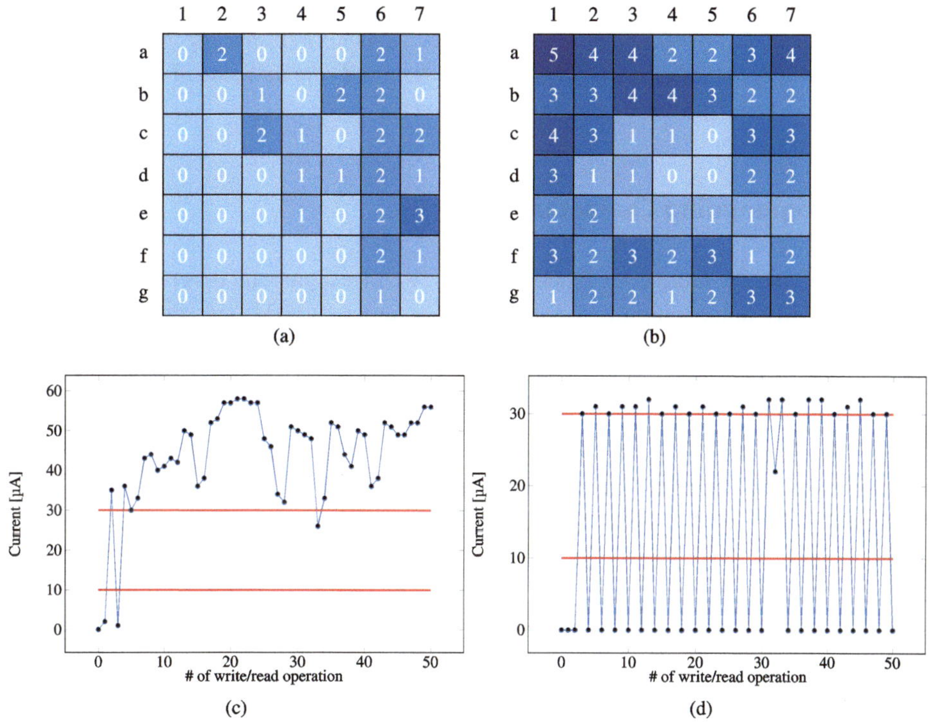

Fig. 6.8 Analysis of the manufacturing yield: heatmap depicting the number of dysfunctional memory cells (**a**) prior to and (**b**) following the forming process across 5 chips. Comparison of switching characteristics between a (**c**) dysfunctional and (**d**) functional memory cell

before and after forming for each device within the 7×7 crossbar array of five chips, employing the IFV algorithm, with parameters detailed in Table 6.1.

Figure 6.8 displays the number of dysfunctional memory cells (a) before and (b) after the forming process. The individual yields of the chips were 85.71, 83.67, 87.76, 81.63, and 95.92%, leading to an average yield of 86.94%. While Milo et al. [16] categorize a cell as formed and functional if it exhibits a read current greater than 18 μA post-forming, we adopt a functional criterion based on a cell's ability to execute at least 50 consecutive switching operations between the HRS and the LRS. For instance, Fig. 6.8 compares the current measurements post-write operation of a dysfunctional (c) and a functional (d) cell. All read operations were performed using a 200 mV read voltage, where a low current indicates the HRS, and a high current indicates the LRS. The dysfunctional cell managed three consecutive switches before it failed, exhibiting a stuck-at LRS fault behavior. Conversely, the functional cell demonstrated stable switching throughout the experiment.

Overall, our findings indicate that a significant number of cells, both before and after forming, are prone to a stuck-at LRS fault. Specifically, we noticed cells with low resistance before forming that did not alter their internal state despite repeated forming attempts. Although correctly formed, some cells could only switch between the HRS and LRS a limited number of times before irreversibly transitioning to the LRS state. In rare instances, this failure occurred after a single reset operation.

6.3.2 Programming Characteristics

After identifying the number of functional cells, we investigate the programming characteristics directly influencing the feasibility and accuracy of CIM operations. To perform a SET or RESET operation, the device interface of the NeuroBreakoutBoard is utilized implementing ISPVA (see Sect. 6.2.2). The respective parameters are given in Table 6.1.

Binary States Initially, the binary switching characteristics is investigated, focusing on the HRS and the LRS3 as the two distinct states. Figure 6.9a–b presents histograms of consecutive alternating SET/RESET operations followed by a read operation for two different cells, highlighting C2C and D2D variability. *While both cells generally exhibit the HRS around 10 μA and the LRS at 30 μA, the cell depicted in Fig. 6.9a demonstrates significant overlap between these states due to two phenomena. Occasionally, although the*

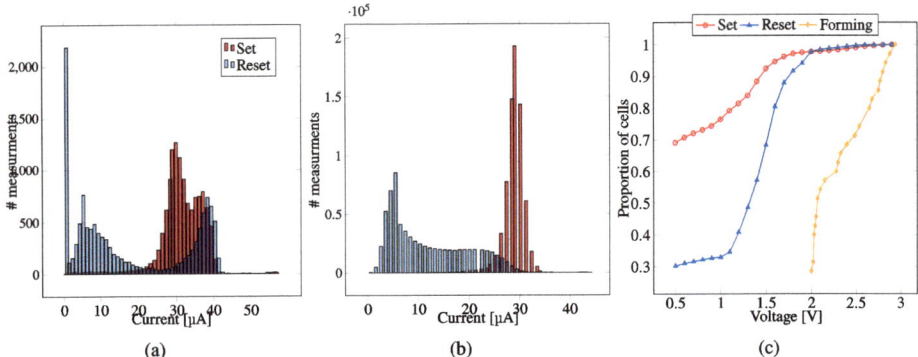

Fig. 6.9 Binary switching characteristics: (**a**)–(**b**) distribution of theHRS and theLRS across two distinct cells. (**c**) CDF analysis for SET, RESET, and forming voltages

RESET operation as indicated by the ISPVA algorithm appears successful, the subsequent read operation inadvertently flips the cell back to its previous state. Additionally, a substantial number of cells display stuck-at faults, leading to insignificant or no changes in resistance following the write algorithm. Overall, variability is more substantial in the HRS compared to the LRS.

Figure 6.9c depicts the CDF of the SET, RESET, and forming voltages for all characterized cells, underscoring the D2D variability observed during programming and forming. *The SET process demonstrates the greatest stability, with over 90% of cells successfully switching at 1.5 V. The CDF for forming voltages highlights the need for an incremental forming algorithm, requiring a forming voltage between 2 V and 3 V to successfully form 90% of the memory cells. The RESET operation, however, exhibits significant variability, ranging from 0.5 V to nearly 3 V, with the lower bound particularly concerning given the typical selection of reading voltages between 0.2 V and 0.5 V.*

Multi-level States   The ability to program memristive devices into multiple levels offers a significant advantage over traditional memory technologies by increasing memory density and enhancing the accuracy of analog vector/matrix multiplication.

The memristive device under study can be programmed into four states (HRS, LRS1, LRS2, and LRS3) by adjusting I_{Target}, as detailed in Table 6.1. However, the inherent variability in memristive devices poses a challenge to multi-level programming by causing state overlaps, making them indistinguishable. To assess the feasibility of achieving distinct states with the provided devices, we cycled the devices through each state 50 times to examine the impact of variability on the different states. Figure 6.10a shows the switching characteristics over 300 programming cycles, with red lines marking I_{Target} for each state. *Generally, LRS1 exhibits considerable variability, sometimes nearly merging with the HRS, attributed to the narrow gap between HRS, defined at a maximum current of 5 μA, and LRS1, set at 10 μA. By contrast, LRS3 shows notably less variation, establishing it as the most stable state, which is why it was selected for binary operation. Figure 6.10b features a histogram of the four states for all measured devices, reinforcing the observation that LRS1 and LRS2 exhibit higher variability compared to LRS3 and HRS.*

6.3.3   Endurance Characteristics

While the programming characteristics impact the precision of operations executed on memristive crossbar arrays, the endurance determines the lifespan of a device, or the number of operations it can perform before failure. Hence, in this section, we delved into the switching endurance of the provided memristive devices, with a focus on contrasting the behaviors of two specific memory cells.

Figure 6.11a displays the programming voltage from the ISPVA for the first device. *Overall, the device sustained approximately 50,000 cycles, equally divided among 12,500 SET and RESET operations, with 25,000 intermediate read operations. A sudden increase in the read current indicates the device's failure, becoming permanently stuck in the LRS.*

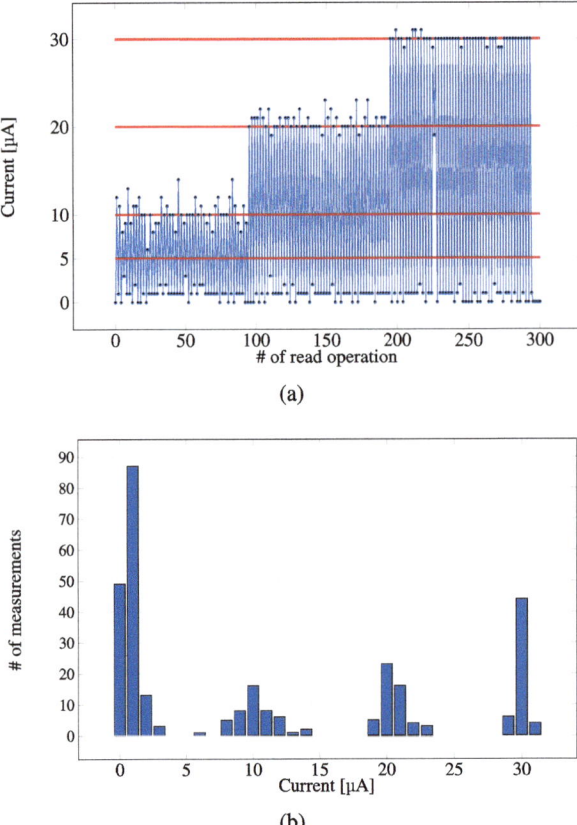

Fig. 6.10 Multi-level switching characteristics: (**a**) repetitive cycling through all applicable resistance states and (**b**) histogram depicting the variability of states

To offer a detailed examination of the device behavior over time, Fig. 6.11b and c depict the cell state after SET (LRS) and RESET (HRS) operations. *While the LRS exhibits reasonable stability, the HRS shows increased variability.* Typically, for the HRS to be effectively distinguishable from the LRS, it should be below 5 µA. *However, for this cell, the HRS frequently overlaps with the LRS, complicating its utility for data storage and CIM operations.*

In comparison, the second cell significantly exceeds the first in terms of cycle count, nearly by two orders of magnitude, as illustrated in Fig. 6.11d. Similar to the first, the LRS of this second cell displays high stability, but fluctuations in the HRS compromise the device's suitability for both memory and CIM applications. Notably, the RESET voltage of the second cell, as shown in Fig. 6.11d, seems more consistent than that of the first cell, suggesting that the instability lies within the HRS, which unintentionally performs a SET operation while its current resistance is being measured.

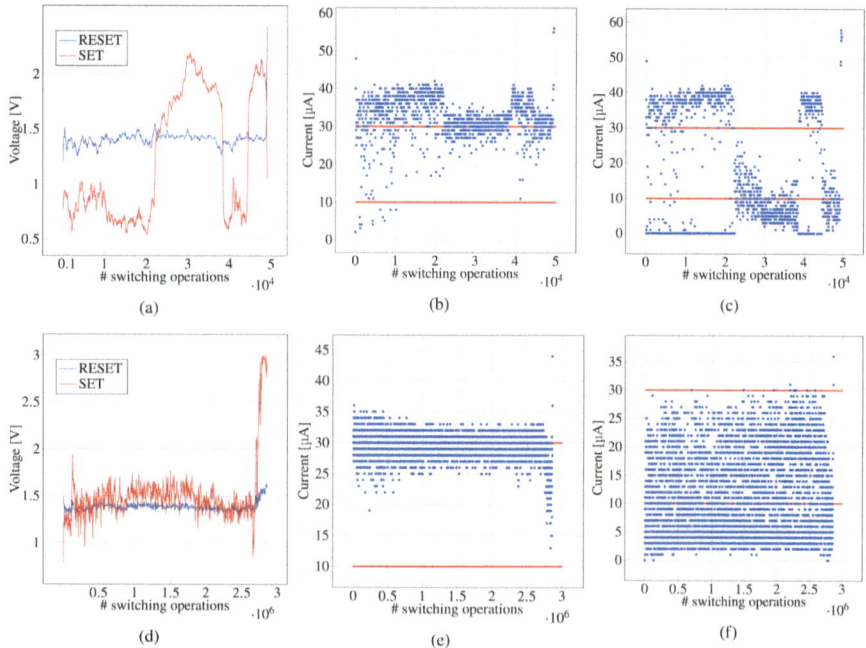

Fig. 6.11 Switching endurance of two memory cells: (**a**)/(**d**) programming voltage for SET/RESET operations and programmed state after (**b**)/(**e**) SET and (**c**)/(**f**) RESET operations

6.3.4   Computing-in-Memory (CIM)

Leveraging its flexible interconnection matrix and signal generation/sensing modules, the NeuroBreakoutBoard excels in executing CIM operations on memristive crossbar arrays. As detailed in Sect. 2.2, CIM operations can be categorized into two types: analog CIM and binary LIM operations. In this section, we evaluate the reliability of both types of operations in the context of the nonidealities inherent in memristive devices.

Analog Computing-in-Memory (CIM)   The analog CIM approach involves performing Multiply–Accumulate (MAC) operations that require an input vector and a corresponding matrix. For this experiment, the input vector consists of four elements, and the matrix dimensions are set to 4×7, constrained by the number of functional cells. Throughout the experiment, we dynamically reprogram the matrix values, represented by the resistance of the memristive devices, with random values. Meanwhile, the input vector is fixed at $\mathbf{v}_{in} = \left(1\ 1\ 1\ 1\right)^{T}$ to ensure comprehensive coverage of all possible output values. Generally, a binary MAC operation involves performing an XNOR operation between the input vector and each matrix column, followed by a Popcount (PC) function that counts the number of 1s in the result vector, representing the MAC operation's outcome.

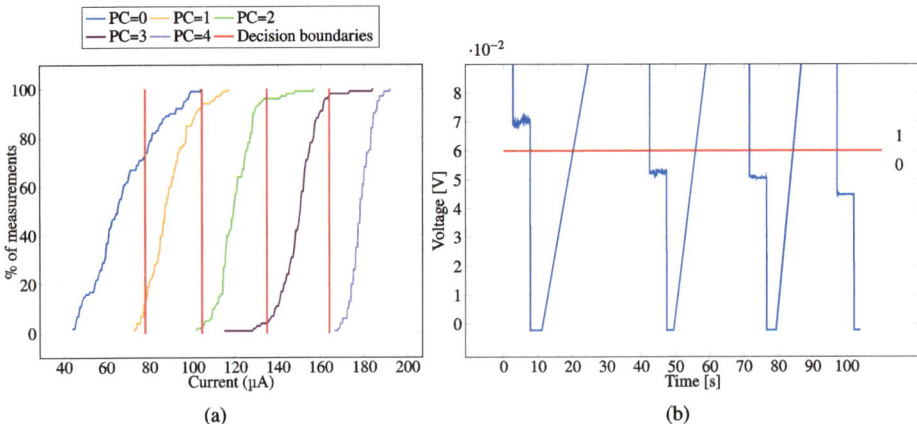

Fig. 6.12 Resilience ofCIM operations: (**a**)CDF of analogCIM operations and (**b**) digital NOR operation executed on the NeuroBreakoutBoard

Figure 6.12a displays the CDF of the accumulated bit line currents, corresponding to the respective popcount values. Given that the input vector comprises four elements, the resulting popcount ranges from 0 (indicating all elements are 0s) to 4 (all elements are 1s). The decision boundaries for these counts are marked with red lines, while the colored traces represent the four distinct popcount outcomes derived from a total of 500 MAC operations. *Overall, the data suggests that the accuracy of the measured accumulated current aligns more closely with the correct result as the popcount increases. For example, 100% of the instances with a popcount of 4 were accurate, whereas for popcounts of 2 and 3, over 90% of the results were correct. However, the scenarios with the lowest popcount (indicating all elements are 0s) exhibited the highest incidence of errors. This finding is consistent with our previous experiments that indicated a higher error susceptibility of the HRS compared to the LRS.*

Binary Logic-in-Memory (LIM)   For executing binary logic operations on memristive crossbar arrays, various logic families have been introduced, outlining the blueprints for constructing logic gates (refer to Sect. 2.2). Initially, our goal was to implement Memristor-Aided Logic (MAGIC) and Memristor-Based Material Implication (IMPLY) gates on the provided crossbar array. However, the effectiveness of both logic families is significantly influenced by specific device characteristics, particularly requiring a dedicated ratio of $\frac{R_{OFF}}{R_{ON}}$ to the memristor's voltage thresholds $V_{T,OFF}$ and $V_{T,ON}$, as detailed in Eq. (2.2). Unfortunately, the memristive devices available to us do not meet these criteria, making these logic families impractical for our purposes.

As a result, we opted to explore the reliability of LIM operations using a Memristor Ratioed Logic (MRL) NOR gate [11–13] with its working principle detailed in Sect. 2.2. Figure 6.12b demonstrates the output voltage of the NOR gate in response to various input combinations $\{(0, 0), (0, 1), (1, 0), (1, 1)\}$, given a computation voltage of 0.2 V. The red line indicates the threshold used to differentiate between logical 1 and 0. *The findings validate the NOR gate's functionality, indicating an output of 0 exclusively for the input pair* (0, 0). *Moreover, the gap between the different output voltages is approximately* 2 mV, *signifying a robust decision margin, especially when considering the significant variability associated with the HRS.*

6.4   Limitations and Outlook

The NeuroBreakoutBoard serves as a flexible and versatile instrumentation platform designed to investigate the impact of memristive devices' inherent characteristics at the device, crossbar, and operation levels. It consists of hardware modules, firmware that operates on the microcontroller, and application interfaces executed on a host PCs, While these components are intricately linked to ensure the platform's easy use, they also introduce certain limitations. The hardware modules support up to 68 analog signals, with 16 of these being assignable to the signal sensing module. Consequently, the size of the crossbar arrays that can be tested is restricted. This constraint extends to the driver and sensing circuitry's maximum ratings, which limit the total current through a bit line and thus the number of rows that can be utilized in a single MAC operation. The firmware and application interfaces are built on a modular architecture. However, the UART/Protobuf communication protocol significantly restricts the overall performance of the NeuroBreakoutBoard. Endurance testing, requiring a comprehensive series of read/write operations, is particularly affected by this bottleneck, leading to extended testing durations. At first glance, these limitations might seem restrictive for experiments across all levels. Yet, the platform plays a pivotal role by facilitating early integration of real devices into the development of neuromorphic computing systems. Given the unique properties of memristive devices, incorporating real devices early is essential for developing reliable and secure computing platforms for the future. In the future, adding an FPGA to the NeuroBreakoutBoard would not only increase the platform's flexibility but also notable enhance its performance. Moreover, implementing LIM gates would greatly benefit from the capability to switch I/Os to a high impedance state, effectively isolating them from the crossbar array and thus reducing unintentional sneak path currents.

6.5   Synopsis

This chapter focuses on the development and capabilities of the NeuroBreakoutBoard, a versatile platform designed for characterizing and performing CIM operations on memristive crossbar arrays. The NBB's design encompasses hardware modules for signal generation, routing, sensing, and interfaces for the device, crossbar, and operation level. The provided case study showcases the NeuroBreakoutBoard's application in evaluating the reliability of a commercially available ReRAM technology, focusing on the manufacturing yield, the programming characteristics, and the execution of CIM operations. The chapter concludes by acknowledging the platform's limitations while emphasizing its crucial role in advancing neuromorphic computing through early real-device integration and potential future improvements.

References

1. Analog Devices Inc.: AD7606 - 8-channel DAS with 16-bit, Bipolar Input, Simultaneous Sampling ADC (2024). https://www.analog.com/media/en/technical-documentation/data-sheets/ad7606_7606-6_7606-4.pdf. Accessed Jan 20, 2024
2. Analog Devices Inc.: AD826 - Low Cost, High Speed, Low Power Dual Operational Amplifier (2024). https://www.analog.com/media/en/technical-documentation/data-sheets/AD826.pdf. Accessed Jan 20, 2024
3. Analog Devices Inc.: ADG1408 - iCMOS Multiplexers (2024). https://www.analog.com/media/en/technical-documentation/data-sheets/adg1408_1409.pdf. Accessed Jan 20, 2024
4. Analog Devices Inc.: ADG613 - 1 pC Charge Injection, 100 pA Leakage, CMOS, ±5 V, +5 V, +3 V, Quad SPST Switches (2024). https://www.analog.com/media/en/technical-documentation/data-sheets/ADG611_612_613.pdf. Accessed Jan 20, 2024
5. Analog Devices Inc.: LT1963A - 1.5A, Low Noise, Fast Transient Response LDO Regulators (2024). https://www.analog.com/media/en/technical-documentation/data-sheets/1963aff.pdf. Accessed Jan 20, 2024
6. Analog Devices Inc.: LT3015 - 1.5A, Low Noise, Negative Linear Regulator with Precision Current Limit (2024). https://www.analog.com/media/en/technical-documentation/data-sheets/3015fb.pdf. Accessed Jan 20, 2024
7. Analog Devices Inc.: LTC2668 - 16-Channel 16-/12-Bit (2024). https://www.analog.com/media/en/technical-documentation/data-sheets/ltc2668.pdf. Accessed Jan 20, 2024
8. Analog Devices Inc.: LTC6269 - Dual 500MHz Ultra-Low Bias Current FET Input Op Amp (2024). https://www.analog.com/media/en/technical-documentation/data-sheets/62689f.pdf. Accessed Jan 20, 2024
9. Analog Devices Inc.: MAX4564 - Low-Voltage, Dual-Supply, SPDT Analog Switch (2024). https://www.analog.com/media/en/technical-documentation/data-sheets/MAX4564.pdf. Accessed Jan 20, 2024
10. Analog Devices Inc.: MAX6126 - Ultra-High-Precision, Ultra-Low-Noise,Series Voltage Reference (2024). https://www.analog.com/media/en/technical-documentation/data-sheets/MAX6126.pdf. Accessed Jan 20, 2024
11. Escudero, M., Vourkas, I., Rubio, A., Moll, F.: Variability-tolerant memristor-based ratioed logic in crossbar array. In: International Symposium on Nanoscale Architectures, NANOARCH '18. ACM (2018). https://doi.org/10.1145/3232195.3232213
12. Escudero, M., Vourkas, I., Rubio, A., Moll, F.: Memristive logic in crossbar memory arrays: variability-aware design for higher reliability. IEEE Trans. Nanotechnol. **18**, 635–646 (2019). https://doi.org/10.1109/tnano.2019.2923731
13. FernANDez, C., Vourkas, I.: Reliability-aware ratioed logic operations for energy-efficient computational ReRAM. In: International Conference on Very Large Scale Integration (VLSI-SoC). IEEE (2022). https://doi.org/10.1109/vlsi-soc54400.2022.9939627
14. Google Inc.: Protocol Buffers Documentation (2024). https://protobuf.dev/. Accessed Jan 24, 2024

15. Grossi, A., Zambelli, C., Olivo, P., MirANDa, E., Stikanov, V., Walczyk, C., Wenger, C.: Electrical characterization and modeling of pulse-based forming techniques in RRAM arrays. Solid-State Electron. **115**, 17–25 (2016). https://doi.org/10.1016/j.sse.2015.10.003

16. Milo, V., Zambelli, C., Olivo, P., Pérez, E., K. Mahadevaiah, M., G. Ossorio, O., Wenger, C., Ielmini, D.: Multilevel HfO2-based rram devices for low-power neuromorphic networks. APL Mater. **7**(8) (2019). https://doi.org/10.1063/1.5108650

17. Pechmann, S., Mai, T., Völkel, M., Mahadevaiah, M.K., Perez, E., Perez-Bosch Quesada, E., Reichenbach, M., Wenger, C., Hagelauer, A.: A versatile, voltage-pulse based read and programming circuit for multi-level RRAM cells. Electronics **10**(5), 530 (2021). https://doi.org/10.3390/electronics10050530

18. Perez, E., Grossi, A., Zambelli, C., Olivo, P., Roelofs, R., Wenger, C.: Reduction of the cell-to-cell variability in Hf1-xAlxOyBased RRAM arrays by using program algorithms. IEEE Electron Device Lett. **38**(2), 175–178 (2017). https://doi.org/10.1109/led.2016.2646758

19. Staudigl, F., Hossein, M., Ziegler, T., Al Indari, H., Pelke, R., Siegel, S., Wouters, D.J., Sisejkovic, D., Joseph, J.M., Leupers, R.: Work-in-progress: a universal instrumentation platform for non-volatile memories. In: International Conference on Hardware/Software Codesign and System Synthesis, CODES/ISSS'23 Companion. ACM (2023). https://doi.org/10.1145/3607888.3608608

20. STMicroelectronics, N.: STM32H745ZI - High-performance AND DSP with DP-FPU, Dual core Arm Cortex-M7+ Cortex-M4 MCU with 2MBytes of Flash Memory, 1MB RAM, 480 MHz CPU, Art Accelerator, L1 Cache, External Memory Interface, Large Set of Peripherals, SMPS (2024). https://www.st.com/resource/en/datasheet/stm32h745zi.pdf. Accessed Jan 20, 2024

21. Traco Power North America Inc.: THL 25-2423 (2024). https://www.tracopower.com/sites/default/files/products/datasheets/thl25_datasheet.pdf. Accessed Jan 20, 2024

22. Traco Power North America Inc.: TMR 3-2421 (2024). y https://www.tracopower.com/sites/default/files/products/datasheets/tmr3_datasheet.pdff. Accessed Jan 20, 2024

23. XP Power Ltd.: JCH1024 (2024). https://www.xppower.com/portals/0/pdfs/SF_JCH10.pdf. Accessed Jan 20, 2024

Conclusion

7

Neuromorphic computing emerges as a promising solution to the von Neumann bottle-neck, which constrains the computing performance and energy efficiency of traditional computing systems. However, the realization of neuromorphic computing hinges on the development of novel devices that can seamlessly integrate computing and data storage within the same location, embodying the computing-in-memory (CIM) paradigm. Memristors, recognized as the fourth fundamental circuit element, stand as the cornerstone for Computing-in-Memory (CIM) yet grapple with reliability issues due to their immature development stage. To overcome the von Neumann bottleneck, neuromorphic computing must demonstrate its efficacy and dependability, ensuring the creation of trustworthy computing systems for future applications.

Therefore, this book sought to investigate the reliability concerns associated with neuromorphic systems, specifically targeting memristor-based CIM architectures.

At the heart of this exploration lies the introduced fault injection platform designed to evaluate the resilience of Logic-in-Memory (LIM) operations to various in-field faults. This platform serves as an essential instrument for examining the effects of imperfections in memristor-based systems and formulating methods to bolster their reliability. The exhaustive evaluation facilitated by this platform illuminated the vulnerabilities of LIM operations to faults, revealing that a certain threshold of in-field faults could be withstood. Notably, the findings indicated that stuck-at faults more severely impact the reliability of emergent applications than bit-flip faults. The presented insights underscore the necessity of implementing fault-tolerant strategies alongside measures to monitor and mitigate the degradation of the memristive devices throughout their lifespan.

Beyond the reliability issues of memristive devices, their distinctive features unveil a new frontier for hardware security threats. NeuroHammer, a significant hardware security attack, leverages these unique attributes to undermine the integrity of computing systems.

© The Author(s), under exclusive license to Springer Nature Switzerland AG 2026   119
F. Staudigl, R. Leupers, *Towards Trustworthy Neuromorphic Computing*,
Synthesis Lectures on Engineering, Science, and Technology,
https://doi.org/10.1007/978-3-032-09586-2_7

Specifically, this attack takes advantage on the high memory density to induce deliberate bit-flips via thermal crosstalk in neighboring memory cells. The provided case study underscores the imperative need for countermeasures to shield against such hardware security threats, advocating for a holistic approach to system design that prioritizes security and reliability from the beginning.

Lastly, the NeuroBreakoutBoard (NBB) is presented as a pivotal platform for the detailed examination of memristive devices from both a reliability and hardware security standpoint. Through enabling the investigation of device characteristics across multiple dimensions, the NeuroBreakoutBoard (NBB) underscores the importance of understanding the underlying mechanisms of memristive behavior to guide the design of more reliable and secure neuromorphic systems.

Based on the presented results, several key requirements can be identified to ensure dependable and secure neuromorphic computing platforms. First, reliability must be addressed across the entire stack, from the device level to the system architecture. While improvements in memristive devices are expected over time, certain non-idealities will remain, requiring solutions at higher abstraction levels. Second, the application itself has a significant impact on reliability. Our results demonstrate that the specific application being executed can accelerate wear on memristive devices, increasing the fault rate. Lastly, hardware security must be a priority during the design phase. Attacks like NeuroHammer, which exploit internal malfunctions, need to be mitigated early in the design process to prevent serious security vulnerabilities.

Conclusively, this book not only addresses the critical challenges of neuromorphic computing but also paves the way for future research directions. By emphasizing the significance of reliability and hardware security in the evolution of memristive-based systems, this work lays the groundwork for the essential progress required to develop trustworthy neuromorphic computing solutions. In addition to the improvement points outlined throughout the book, the following research directions show promise:

- Advanced Error Mitigation Techniques: Error Correcting Codes (ECCs) are a crucial part of modern memories for detecting and mitigating the impact of faults. These well-established techniques cannot simply be applied to Resistive Random-Access Memories (ReRAMs) due to their use of CIM. Therefore, novel ECCs must be developed that account for both memory and CIM applications, while also considering device wear.
- Reliability-aware Mapping Algorithms: Mapping algorithms are a crucial piece of software that assign a given neural network to a specific system architecture. While these algorithms already take into account factors like data flow and the number of compute cores, memristor-based accelerators could benefit from reliability-aware mapping. Such algorithms could optimize device usage and data flow to significantly reduce wear over time.
- Security-Enhanced Architectures: As neuromorphic computing systems continue to evolve, addressing security becomes critical to defend against advanced attacks like

NeuroHammer and other exploits. Mitigation strategies must carefully balance the trade-offs between area, power, and performance overhead against the security benefits. Ensuring this balance is essential to maintaining the overall performance and efficiency of future neuromorphic systems while providing robust security protections. Research should focus on developing lightweight, efficient security mechanisms that can be integrated without compromising system resources or computational capabilities.

Simulation Details

A

This appendix provides detailed information on the JART simulation model discussed in Chap. 5. Additionally, it includes documentation of the alpha matrices essential for conducting circuit-level simulations to verify the NeuroHammer attack (Tables A.1, A.2, A.3, A.4, A.5, A.6, A.7, A.8, A.9, A.10, and A.11).

A.1   Model Parameter

Table A.1 JART VCM v1b model parameters

$A_{det} = \pi r^2 = 6.36 \times 10^{-15}\,\mathrm{m}^2$	$r_{det} = 45\,\mathrm{nm}$
$N_{disc,min,det} = 0.008 \times 10^{26}\,\mathrm{m}^{-3}$	$N_{disc,max,det} = 20 \times 10^{26}\,\mathrm{m}^{-3}$
$l_{cell} = 3\,\mathrm{nm}$	$l_{det} = 0.4\,\mathrm{nm}$
$l_{plug} = 2.6\,\mathrm{nm}$	$a = 0.25\,\mathrm{nm}$
$R_{series} = 1.37\,\mathrm{k\Omega}\ (I = 0\,\mu\mathrm{A})$	$R_{series} = 1.46\,\mathrm{k\Omega}\ (I = 700\,\mu\mathrm{A})$
$R_{line} = 719\,\Omega\ (I = 0\,\mu\mathrm{A})$	$R_{line} = 810\,\Omega\ (I = 700\,\mu\mathrm{A})$
$R_{TiOx} = 650\,\Omega$	$\Delta W_A = 1.35\,\mathrm{eV}$
$\nu_0 = 2 \times 10^{13}\,\mathrm{Hz}$	$\mu_n = 4 \times 10^{-6}\,\mathrm{m2/(V\ s)}$
$R_{th0,SET} = 15.72 \times 10^6\,\mathrm{K/W}$	$R_{th0,RESET} = 4.24 \times 10^6\,\mathrm{K/W}$
$e\phi_{n0} = 0.18\,\mathrm{eV}$	$e\phi_n = 0.1\,\mathrm{eV}$
$R_{th,line} = 90\,471.47\,\mathrm{K/W}$	$R_0 = 719.24\,\Omega$
$\alpha_{line} = 3.92 \times 10^{-3}\,/\mathrm{K}$	$m^* = 9.11 \times 10^{-31}\,\mathrm{kg}$
$e = 1.6 \times 10^{-19}\,\mathrm{C}$	$T_0 = 293\,\mathrm{K}$
$A^* = 6.01 \times 10^5\,\mathrm{A/(m2\ K2)}$	$z_{V_O} = 2$
$k_B = 1.38 \times 10^{-23}\,\mathrm{J/K}$	$\varepsilon_0 = 8.854 \times 10^{-12}\,\mathrm{A\ s/(V\ m)}$
$\varepsilon_{\phi_B} = 5.5\varepsilon_0$	$\varepsilon = 17\varepsilon_0$
$h = 6.626 \times 10^{-34}\,\mathrm{J\ s}$	

© The Author(s), under exclusive license to Springer Nature Switzerland AG 2026
F. Staudigl, R. Leupers, *Towards Trustworthy Neuromorphic Computing*,
Synthesis Lectures on Engineering, Science, and Technology,
https://doi.org/10.1007/978-3-032-09586-2

A.2   Alpha Matrices

Table A.2 9×9 alpha matrix determined for a 10 nm electrode spacing

0.0233	0.0285	0.0333	0.0369	0.0383	0.0369	0.0333	0.0285	0.0233
0.0290	0.0365	0.0442	0.0505	0.0532	0.0505	0.0442	0.0366	0.0291
0.0357	0.0466	0.0597	0.0729	0.0801	0.0729	0.0598	0.0467	0.0358
0.0417	0.0573	0.0791	0.1084	0.1397	0.1085	0.0791	0.0574	0.0419
0.0444	0.0628	0.0930	0.1519	1.0000	0.1520	0.0932	0.0630	0.0446
0.0416	0.0572	0.0789	0.1082	0.1392	0.1082	0.0790	0.0573	0.0418
0.0355	0.0464	0.0594	0.0725	0.0794	0.0725	0.0594	0.0465	0.0356
0.0288	0.0362	0.0438	0.0499	0.0523	0.0499	0.0438	0.0363	0.0289
0.0231	0.0282	0.0328	0.0362	0.0373	0.0362	0.0329	0.0282	0.0231

Table A.3 9×9 alpha matrix determined for a 20 nm electrode spacing

0.0158	0.0201	0.0241	0.0270	0.0283	0.0271	0.0241	0.0201	0.0158
0.0212	0.0278	0.0346	0.0402	0.0427	0.0403	0.0346	0.0279	0.0213
0.0278	0.0380	0.0505	0.0631	0.0700	0.0632	0.0506	0.0381	0.0279
0.0341	0.0494	0.0714	0.1016	0.1324	0.1017	0.0715	0.0496	0.0343
0.0370	0.0557	0.0879	0.1526	1.0000	0.1528	0.0881	0.0560	0.0373
0.0340	0.0493	0.0712	0.1013	0.1317	0.1014	0.0713	0.0494	0.0342
0.0276	0.0378	0.0501	0.0625	0.0690	0.0626	0.0502	0.0379	0.0278
0.0210	0.0275	0.0341	0.0395	0.0416	0.0395	0.0341	0.0275	0.0210
0.0156	0.0197	0.0235	0.0262	0.0270	0.0262	0.0236	0.0198	0.0156

Table A.4 9×9 alpha matrix determined for a 30 nm electrode spacing

0.0108	0.0143	0.0176	0.0201	0.0212	0.0202	0.0177	0.0144	0.0109
0.0156	0.0213	0.0273	0.0325	0.0347	0.0325	0.0274	0.0213	0.0156
0.0216	0.0310	0.0427	0.0549	0.0615	0.0549	0.0428	0.0311	0.0217
0.0276	0.0423	0.0641	0.0949	0.1257	0.0950	0.0642	0.0425	0.0278
0.0304	0.0488	0.0819	0.1510	1.0000	0.1512	0.0822	0.0491	0.0308
0.0275	0.0421	0.0638	0.0945	0.1249	0.0946	0.0640	0.0423	0.0277
0.0214	0.0307	0.0423	0.0542	0.0603	0.0543	0.0424	0.0308	0.0215
0.0154	0.0210	0.0268	0.0316	0.0333	0.0316	0.0268	0.0210	0.0154
0.0106	0.0140	0.0171	0.0192	0.0197	0.0192	0.0171	0.0140	0.0107

Table A.5 9×9 alpha matrix determined for a 40 nm electrode spacing

0.0074	0.0102	0.0129	0.0149	0.0158	0.0150	0.0129	0.0102	0.0074
0.0113	0.0162	0.0215	0.0260	0.0281	0.0261	0.0215	0.0162	0.0114
0.0166	0.0250	0.0358	0.0474	0.0538	0.0475	0.0359	0.0251	0.0167
0.0220	0.0357	0.0568	0.0877	0.1185	0.0878	0.0570	0.0359	0.0222
0.0246	0.0421	0.0751	0.1473	1.0000	0.1476	0.0755	0.0425	0.0250
0.0219	0.0355	0.0566	0.0872	0.1175	0.0873	0.0567	0.0357	0.0221
0.0164	0.0247	0.0354	0.0466	0.0523	0.0466	0.0355	0.0248	0.0165
0.0111	0.0158	0.0209	0.0251	0.0265	0.0251	0.0209	0.0159	0.0112
0.0072	0.0098	0.0123	0.0139	0.0141	0.0139	0.0123	0.0098	0.0072

Table A.6 9×9 alpha matrix determined for a 50 nm electrode spacing

0.0050	0.0071	0.0092	0.0109	0.0117	0.0109	0.0093	0.0071	0.0050
0.0081	0.0121	0.0166	0.0206	0.0224	0.0206	0.0166	0.0122	0.0082
0.0125	0.0198	0.0296	0.0403	0.0464	0.0404	0.0297	0.0199	0.0126
0.0172	0.0295	0.0496	0.0798	0.1103	0.0799	0.0498	0.0298	0.0174
0.0194	0.0356	0.0677	0.1412	1.0000	0.1415	0.0681	0.0361	0.0200
0.0171	0.0294	0.0493	0.0792	0.1090	0.0794	0.0495	0.0296	0.0173
0.0123	0.0195	0.0291	0.0394	0.0447	0.0395	0.0292	0.0196	0.0124
0.0079	0.0118	0.0160	0.0195	0.0206	0.0196	0.0160	0.0118	0.0080
0.0048	0.0068	0.0086	0.0098	0.0097	0.0098	0.0086	0.0068	0.0048

Table A.7 9×9 alpha matrix determined for a 60 nm electrode spacing

0.0033	0.0049	0.0066	0.0080	0.0086	0.0080	0.0066	0.0050	0.0033
0.0058	0.0090	0.0128	0.0162	0.0179	0.0163	0.0128	0.0091	0.0058
0.0093	0.0156	0.0243	0.0342	0.0399	0.0343	0.0244	0.0157	0.0095
0.0133	0.0243	0.0430	0.0723	0.1023	0.0725	0.0432	0.0245	0.0136
0.0152	0.0298	0.0606	0.1346	1.0000	0.1350	0.0611	0.0304	0.0158
0.0132	0.0241	0.0426	0.0717	0.1009	0.0718	0.0429	0.0244	0.0135
0.0092	0.0153	0.0238	0.0332	0.0380	0.0333	0.0239	0.0154	0.0093
0.0056	0.0087	0.0122	0.0151	0.0158	0.0151	0.0122	0.0087	0.0056
0.0032	0.0046	0.0060	0.0068	0.0064	0.0068	0.0060	0.0046	0.0032

Table A.8 9 × 9 alpha matrix determined for a 70 nm electrode spacing

0.0022	0.0034	0.0047	0.0058	0.0064	0.0058	0.0047	0.0034	0.0022
0.0041	0.0067	0.0098	0.0128	0.0142	0.0128	0.0098	0.0067	0.0041
0.0069	0.0122	0.0199	0.0289	0.0342	0.0290	0.0200	0.0123	0.0071
0.0102	0.0199	0.0372	0.0655	0.0950	0.0656	0.0374	0.0202	0.0105
0.0118	0.0249	0.0541	0.1282	1.0000	0.1286	0.0547	0.0256	0.0125
0.0102	0.0197	0.0368	0.0647	0.0933	0.0649	0.0371	0.0200	0.0104
0.0068	0.0119	0.0194	0.0279	0.0321	0.0279	0.0195	0.0121	0.0069
0.0039	0.0064	0.0092	0.0116	0.0119	0.0116	0.0092	0.0064	0.0040
0.0021	0.0031	0.0041	0.0046	0.0040	0.0046	0.0041	0.0031	0.0021

Table A.9 9 × 9 alpha matrix determined for a 80 nm electrode spacing

0.0015	0.0024	0.0034	0.0043	0.0048	0.0043	0.0035	0.0024	0.0015
0.0029	0.0050	0.0077	0.0102	0.0115	0.0103	0.0077	0.0051	0.0030
0.0052	0.0097	0.0166	0.0249	0.0300	0.0250	0.0167	0.0098	0.0053
0.0080	0.0165	0.0326	0.0602	0.0897	0.0604	0.0329	0.0168	0.0083
0.0092	0.0210	0.0490	0.1240	1.0000	0.1245	0.0497	0.0218	0.0100
0.0079	0.0163	0.0323	0.0594	0.0877	0.0596	0.0325	0.0166	0.0082
0.0051	0.0095	0.0160	0.0238	0.0275	0.0238	0.0161	0.0096	0.0052
0.0028	0.0047	0.0070	0.0089	0.0089	0.0090	0.0071	0.0048	0.0028
0.0014	0.0021	0.0028	0.0030	0.0021	0.0030	0.0028	0.0022	0.0014

Table A.10 9 × 9 alpha matrix determined for a 90 nm electrode spacing

0.00097	0.00164	0.00240	0.00308	0.00351	0.00309	0.00241	0.00165	0.00098
0.00202	0.00362	0.00574	0.00788	0.00899	0.00790	0.00577	0.00367	0.00205
0.00378	0.00740	0.01321	0.02059	0.02516	0.02065	0.01330	0.00752	0.00388
0.00592	0.01310	0.02745	0.05317	0.08136	0.05334	0.02774	0.01341	0.00621
0.00680	0.01697	0.04253	0.11498	1.0000	0.11558	0.04332	0.01783	0.00764
0.00586	0.01294	0.02709	0.05232	0.07916	0.05251	0.02738	0.01326	0.00614
0.00367	0.00715	0.01264	0.01937	0.02245	0.01943	0.01274	0.00727	0.00377
0.00189	0.00333	0.00510	0.00651	0.00614	0.00652	0.00513	0.00337	0.00193
0.00086	0.00138	0.00181	0.00172	0.00057	0.00172	0.00181	0.00139	0.00087

Table A.11 The 1×32 alpha matrix is utilized for simulating 1T1R structures. For visualization purposes, the line vector is presented across multiple rows

0.2463	0.2381	0.2234	0.2069	0.1909	0.1759	0.1623	0.1501
0.1394	0.1303	0.1234	0.1205	0.1301	0.1652	0.2638	0.5242
1.0000	0.4763	0.1871	0.0743	0.0301	0.0127	0.0059	0.0031
0.0020	0.0016	0.0013	0.0012	0.0012	0.0011	0.0010	0.0009

Communication Details

B

This appendix provides detailed information on the defined Protocol Buffers messages used to establish the communication between the host and the STM32 microcontroller (Figs. B.1, B.2, and B.3).

Fig. B.1 Protocol buffers file specifying the message structure to determine the operational mode of the NeuroBreakoutBoard

```
enum _Operation{
    NONE           = 0;
    read           = 1;
    form           = 2;
    set            = 3;
    reset          = 4;
    multiplication = 5;
    readRow        = 6;
    NOR            = 7;
    NOT            = 8;
}
```

© The Author(s), under exclusive license to Springer Nature Switzerland AG 2026
F. Staudigl, R. Leupers, *Towards Trustworthy Neuromorphic Computing*,
Synthesis Lectures on Engineering, Science, and Technology,
https://doi.org/10.1007/978-3-032-09586-2

```
message Response{
        _Operation operation = 1;
        uint32 bitLine      = 2; // Bitline index of the target
   cell
        uint32 sourceLine   = 3; // Sourceline index of the target
    cell
        uint32 wordLine     = 4; // Wordline index of the target
   cell
        uint32 resistance   = 5; // Measured resistance of the
   cell
        int32 vSet          = 6; // Final programming voltage (
   ISPVA)
        repeated int32 currentList = 7; // List of all measured
   currents
}
```

Fig. B.2 Protocol buffers file defining the response message sent from the STM32 microcontroller to the host PCs

```
message Request{
 _Operation operation  = 1;
 uint32 bitLine    = 2; // Bitline index of the target cell
 uint32 sourceLine = 3; // Sourceliune index of the target cell
 uint32 wordLine   = 4; // Wordline index of the target cell
 float vSet     = 5; // Start voltage of the ISPVA [V]
 float vStop    = 6; // Stop voltage of the ISPVA [V]
 float vStep    = 7; // Voltage difference in every step of ISPVA [
    V]
 float tStep    = 8; // Pulse duration of the ISPVA [us]
 float vGate    = 9; // Gate voltage for the set/form operation [V]
 float vRead    = 10; // Read voltage for the verify step in ISPVA
    [V]
 float iTarget    = 11; // Target current of the ISPVA [uA]
 float vGateReset = 12; // Gate voltage for the reset operation [V
    ]
 uint32 vBL_0   = 13; // Binary input vector for multiplication
 uint32 vBL_1   = 14;
 uint32 vBL_2   = 15;
 uint32 vBL_3   = 16;
 uint32 vBL_4   = 17;
 uint32 vBL_5   = 18;
 uint32 vBL_6   = 19;
 uint32 vHigh   = 20; // Voltage representing logic 1
 uint32 vLow    = 21; // Voltage representing logic 0
 float v0    = 22; // Computation voltage of the logic gate
 uint32 logicBL0   = 23; // 1st bitline index of the logic gate
 uint32 logicSL0   = 24; // 2nd bitline index of the logic gate
 uint32 logicSL1   = 25; // Sourceline index of the logic gate
 float vGateRead   = 26; // Gate voltage for the verify step in
    ISPVA[V]
 }
```

Fig. B.3 Protocol buffers file specifying the request message sent from the host PCs to the STM32 microcontroller

Index

© The Author(s), under exclusive license to Springer Nature Switzerland AG 2026
F. Staudigl, R. Leupers, *Towards Trustworthy Neuromorphic Computing*,
Synthesis Lectures on Engineering, Science, and Technology,
https://doi.org/10.1007/978-3-032-09586-2

MIX
Papier aus verantwortungsvollen Quellen
Paper from responsible sources
FSC® C105338

If you have any concerns about our products,
you can contact us on
ProductSafety@springernature.com

In case Publisher is established outside the EU,
the EU authorized representative is:
Springer Nature Customer Service Center GmbH
Europaplatz 3, 69115 Heidelberg, Germany

Printed by Libri Plureos GmbH
in Hamburg, Germany